江西理工大学资助出版

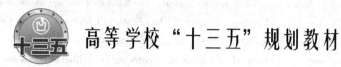

高等学校"十三五"规划教材

Introduction of Physical Mineral Processing

Xinyang Yu

Zhiqiang Huang   Hui Wang   Guichun He

Beijing

Metallurgical Industry Press

2018

Summary

This book presents an introduction to the basic concepts, process methods and equipment of physical mineral processing. It is composed of seven chapters. The main contents are as follows: Chapter 1 Introduction, Chapter 2 Ore Handling, Chapter 3 Classification, Chapter 4 Gravity Concentration, Chapter 5 Dense Medium Separation (DMS), Chapter 6 Magnetic and Electrical Separation, and Chapter 7 Dewatering. It can be used as a textbook to carry out the bilingual education and international cooperation education of mineral processing engineering, or other related fields of the colleges, institutions and universities. It also can be used to improve the Academic English for the researcher working in mineral processing.

图书在版编目(CIP)数据

矿物加工概论=Introduction of Physical Mineral Processing：英文/
余新阳等编著． —北京：冶金工业出版社，2018.12
高等学校"十三五"规划教材
ISBN 978-7-5024-8032-5

Ⅰ.①矿… Ⅱ.①余… Ⅲ.①选矿—高等学校—教材—英文
Ⅳ.①TD9

中国版本图书馆 CIP 数据核字(2018)第 301527 号

出 版 人　谭学余
地　　址　北京市东城区嵩祝院北巷 39 号　邮编　100009　电话　(010)64027926
网　　址　www.cnmip.com.cn　电子信箱　yjcbs@cnmip.com.cn
责任编辑　杨盈园　美术编辑　彭子赫　版式设计　孙跃红
责任校对　王永欣　责任印制　李玉山
ISBN 978-7-5024-8032-5
冶金工业出版社出版发行；各地新华书店经销；固安华明印业有限公司印刷
2018 年 12 月第 1 版，2018 年 12 月第 1 次印刷
787mm×1092mm　1/16；16 印张；385 千字；246 页
36.00 元
冶金工业出版社　投稿电话　(010)64027932　投稿信箱　tougao@cnmip.com.cn
冶金工业出版社营销中心　电话　(010)64044283　传真　(010)64027893
冶金工业出版社天猫旗舰店　yjgycbs.tmall.com
(本书如有印装质量问题，本社营销中心负责退换)

前　言

　　矿产资源是人类社会生存与发展的重要物质基础，是一种有限的、不可再生的自然资源。矿物加工工程是矿产资源开发利用的一个重要学科，根据选别矿物的方法与原理不同，主要分为物理选矿、浮游选矿、化学选矿等选矿方法。随着全球经济一体化的高速发展，我国矿业经济和人才培养的对外交流与合作也在不断增加，对矿物加工工程人才更是提出了越来越高的要求。为了适应新形势下人才培养的要求，编者在参考国内外大量文献资料的基础上，结合课题组多年积累的专业知识和研究经验，编写了英文版的物理选矿概论——*Introduction of Physical Mineral Processing*，旨在使矿物加工专业人员通过本书的学习提高专业外语水平。此外，本书还可作为高等学校矿物加工工程及相关专业开展双语教学、国际合作办学的教学用书。

　　本书概述了物理选矿的基本概念、工艺方法及设备等，共7章，主要内容有概述（Chapter 1 Introduction）、原矿处理（Chapter 2 Ore Handling）、分级（Chapter 3 Classification）、重力选矿（Chapter 4 Gravity Concentration）、重介质选矿（Chapter 5 Dense Medium Separation（DMS））、磁电选矿（Chapter 6 Magnetic and Electrical Separation）、脱水（Chapter 7 Dewatering）等。全书由余新阳任主编，黄志强、王晖、何桂春任副主编，余新阳负责全书的统一整理和校核。其中第1章、第4章、第5章由余新阳、王礼平、胡琳琪编写，第2章由何桂春编写，第3章和总词汇（Vocabulary）由黄志强、成晟编写，第6章由何桂春、黄超军、李少平、李坤编写，第7章由王晖、郭腾博编写。

　　本书由江西理工大学资助出版，在此表示衷心的感谢。

　　由于编者水平所限，书中若有疏漏之处，诚望读者批评指正。

编　者

2018年12月

Contents

Chapter 1 Introduction 1

1.1 MINERALS 1
1.2 ABUNDANCE OF MINERALS 3
1.3 DEPOSITS AND ORES 5
1.4 METALLIC AND NONMETALLIC ORES 5
1.5 THE NEED FOR MINERAL PROCESSING 6
1.6 LIBERATION 7
1.7 CONCENTRATION 11
1.8 REPRESENTING MINERAL PROCESSING SYSTEMS:
    THE FLOWSHEET 14
1.9 MEASURES OF SEPARATION 14
    1.9.1 Grade 15
    1.9.2 Recovery 15
    1.9.3 Grade-Recovery Relationship 16
    1.9.4 A Measure of Technical Separation Efficiency 17
1.10 ECONOMIC CONSIDERATIONS 19
    1.10.1 Contained Value 21
    1.10.2 Processing Costs 21
    1.10.3 Milling Costs 24
    1.10.4 Tailings Reprocessing and Recycling 24
    1.10.5 Net Smelter Return and Economic Efficiency 25
    1.10.6 Case Study: Economics of Tin Processing 27
    1.10.7 Case Study: Economics of Copper Processing 31
    1.10.8 Economic Efficiency 35
1.11 SUSTAINABILITY 38
REFERENCES 41

Chapter 2 Ore Handling 44

2.1 INTRODUCTION 44
2.2 THE REMOVAL OF HARMFUL 44
2.3 ORE TRANSPORTATION 47
2.4 ORE STORAGE 53

2.5 FEEDING ········· 55
2.6 SELF-HEATING OF SULFIDE MINERALS ········· 57
REFERENCES ········· 59

Chapter 3 Classification ········· 61

3.1 INTRODUCTION ········· 61
3.2 PRINCIPLES OF CLASSIFICATION ········· 61
   3.2.1 Force Balance ········· 61
   3.2.2 Free Settling ········· 62
   3.2.3 Hindered Settling ········· 64
   3.2.4 Effect of Density on Separation Efficiency ········· 64
   3.2.5 Effect of Classifier Operation on Grinding Circuit Behavior ········· 67
3.3 TYPES OF CLASSIFIERS ········· 67
3.4 CENTRIFUGAL CLASSIFIERS-THE HYDROCYCLONE ········· 68
   3.4.1 Basic Design and Operation ········· 68
   3.4.2 Characterization of Cyclone Efficiency ········· 71
   3.4.3 Hydrocyclones Versus Screens ········· 77
   3.4.4 Mathematical Models of Hydrocyclone Performance ········· 78
   3.4.5 Operating and Geometric Factors Affecting Cyclone Performance ········· 80
   3.4.6 Sizing and Scale-Up of Hydrocyclones ········· 84
3.5 GRAVITATIONAL CLASSIFIERS ········· 89
   3.5.1 Sedimentation Classifiers ········· 89
   3.5.2 Hydraulic Classifiers ········· 93
REFERENCES ········· 94

Chapter 4 Gravity Concentration ········· 97

4.1 INTRODUCTION ········· 97
4.2 PRINCIPLES OF GRAVITY CONCENTRATION ········· 97
4.3 GRAVITATIONAL CONCENTRATORS ········· 98
   4.3.1 Jigs ········· 100
   4.3.2 Spirals ········· 109
   4.3.3 Shaking Tables ········· 112
4.4 CENTRIFUGAL CONCENTRATORS ········· 116
4.5 SLUICES AND CONES ········· 120
4.6 FLUIDIZED BED SEPARATORS ········· 122
4.7 DRY PROCESSING ········· 123
4.8 SINGLE-STAGE UNITS AND CIRCUITS ········· 124
REFERENCES ········· 126

Chapter 5　Dense Medium Separation　130

5.1　INTRODUCTION　130
5.2　THE DENSE MEDIUM　131
　5.2.1　Liquids　131
　5.2.2　Suspensions　132
5.3　SEPARATING VESSELS　133
　5.3.1　Gravitational Vessels　133
　5.3.2　Centrifugal Separators　136
5.4　DMS CIRCUITS　140
5.5　EXAMPLE DMS APPLICATIONS　142
5.6　LABORATORY HEAVY LIQUID TESTS　144
5.7　EFFICIENCY OF DMS　148
REFERENCES　156

Chapter 6　Magnetic and Electrical Separation　159

6.1　INTRODUCTION　159
6.2　MAGNETISM IN MINERALS　159
6.3　EQUATIONS OF MAGNETISM　162
6.4　MAGNETIC SEPARATOR DESIGN　164
　6.4.1　Magnetic Field Gradient　164
　6.4.2　Magnetic Field Intensity　167
　6.4.3　Material Transport in Magnetic Separators　168
6.5　TYPES OF MAGNETIC SEPARATOR　169
　6.5.1　Low-Intensity Magnetic Separators　169
　6.5.2　High-Intensity Magnetic Separators　174
　6.5.3　High-Gradient Magnetic Separators　178
　6.5.4　Superconducting Separators　181
6.6　ELECTRICAL SEPARATION　184
　6.6.1　Ion Bombardment　184
　6.6.2　Conductive Induction　188
　6.6.3　Triboelectric Charging　190
　6.6.4　Example Flowsheets　194
REFERENCES　196

Chapter 7　Dewatering　200

7.1　INTRODUCTION　200
7.2　GRAVITATIONAL SEDIMENTATION　201
　7.2.1　Sedimentation of Particles　201

7.2.2	Particle Aggregation: Coagulation and Flocculation	202
7.2.3	Thickener Types	207
7.2.4	Thickener Operation and Control	211
7.2.5	Thickener Sizing	213
7.2.6	Other Gravity Sedimentation Devices	215

7.3 CENTRIFUGAL SEDIMENTATION ································· 216
7.4 FILTRATION ··· 218
    7.4.1 Brief Theory ·· 218
    7.4.2 The Filter Medium ··· 219
    7.4.3 Filtration Tests ··· 219
    7.4.4 Types of Filter ··· 221
7.5 DRYING ··· 227
REFERENCES ··· 228

Vocabulary ··· 232

Chapter 1 Introduction

1.1 MINERALS

The forms in which metals are found in the crust of the earth and as seabed deposits depend on their reactivity with their environment, particularly with oxygen, sulfur, and carbon dioxide. Gold and platinum metals are found principally in the native or metallic form. Silver, copper, and mercury are found native as well as in the form of sulfides, carbonates, and chlorides. The more reactive metals are always in compound form, such as the oxides and sulfides of iron and the oxides and silicates of aluminum and beryllium. These naturally occurring compounds are known as minerals, most of which have been given names according to their composition (e.g., galena-lead sulfide, PbS; cassiterite-tin oxide, SnO_2).

Minerals by definition are natural inorganic substances possessing definite chemical compositions and atomic structures. Some flexibility, however, is allowed in this definition. Many minerals exhibit isomorphism, where substitution of atoms within the crystal structure by similar atoms takes place without affecting the atomic structure. The mineral olivine, for example, has the chemical composition $(Mg, Fe)_2 SiO_4$, but the ratio of Mg atoms to Fe atoms varies. The total number of Mg and Fe atoms in all olivines, however, has the same ratio to that of the Si and O atoms. Minerals can also exhibit polymorphism, different minerals having the same chemical composition, but markedly different physical properties due to a difference in crystal structure. Thus, the two minerals graphite and diamond have exactly the same composition, being composed entirely of carbon atoms, but have widely different properties due to the arrangement of the carbon atoms within the crystal lattice.

The term "mineral" is often used in a much more extended sense to include anything of economic value that is extracted from the earth. Thus, coal, chalk, clay, and granite do not come within the definition of a mineral, although details of their production are usually included in national figures for mineral production. Such materials are, in fact, rocks, which are not homogeneous in chemical and physical composition, as are minerals, but generally consist of a variety of minerals and form large parts of the earth's crust. For instance, granite, which is one of the most abundant igneous rocks, that is, a rock formed by cooling of molten material, or magma, within the earth's crust, is composed of three main mineral constituents: feldspar, quartz, and mica. These three mineral components occur in varying proportions in different parts of the same granite mass.

Coals are a group of bedded rocks formed by the accumulation of vegetable matter. Most coalseams were formed over 300 million years ago by the decomposition of vegetable matter from the dense tropical forests which covered certain areas of the earth. During the early formation of the coal-seams, the rotting vegetation formed thick beds of peat, an unconsolidated product of the decomposition of vegetation, found in marshes and bogs. This later became overlain with shales, sandstones, mud, and silt, and under the action of the increasing pressure, temperature and time, the peat-beds became altered, or metamorphosed, to produce the sedimentary rock known as coal. The degree of alteration is known as the rank of the coal, with the lowest ranks (lignite or brown coal) showing little alteration, while the highest rank (anthracite) is almost pure graphite (carbon).

While metal content in an ore is typically quoted as percent metal, it is important to remember that the metal is contained in a mineral (e. g. , tin in SnO_2). Depending on the circumstances it may be necessary to convert from metal to mineral, or vice versa. The conversion is illustrated in the following two examples (Examples 1. 1 and 1. 2).

The same element may occur in more than one mineral and the calculation becomes a little more involved.

Example 1. 1

Given a tin concentration of 2. 00% in an ore, what is the concentration of cassiterite (SnO_2)?

Solution

Step 1: What is the Sn content of SnO_2?

Molar mass of Sn (M_{Sn}) 118. 71g/mol; Molar mass of O (M_O) 15. 99g/mol

$$\%Sn \text{ in } SnO_2 = \frac{M_{Sn}}{M_{Sn} + 2 \times M_O} = \frac{118.71}{118.71 + 2 \times 15.99} = 78.8\%$$

Step 2: Convert Sn concentration to SnO_2

$$\frac{2.00\% Sn}{78.8\% Sn \text{ in } SnO_2} = 2.54\% SnO_2$$

Example 1. 2

A sample contains three phases, chalcopyrite ($CuFeS_2$), pyrite (FeS_2), and non-sulfides (containing no Cu or Fe). If the Cu concentration is 22. 5% and the Fe concentration is 25. 6%, what is the concentration of pyrite and of the nonsulfides?

Solution

Note, Fe occurs in two minerals which is the source of complication. The solution, in this case, is to calculate first the % chalcopyrite using the $w(Cu)$ data in a similar manner to the calculation in Example 1. 1 (Step 1), and then to calculate the %Fe contributed by the Fe in the chalcopyrite (Step 2) from which $w(Fe)$ associated with pyrite can be calculated (Step 3).

Molar masses (g/mol): Cu 63. 54; Fe 55. 85; S 32. 06.

Step 1: Convert Cu to chalcopyrite (Cp)

$$w(Cp) = 22.5\% \left[\frac{63.54 + 55.85 + (2 \times 32.06)}{63.54} \right] = 65.0\%$$

Step 2: Determine $w(Fe)$ in Cp

$$w(Fe) \text{ in } Cp = 65\% \left[\frac{55.85}{63.54 + 55.85 + (2 \times 32.06)} \right] = 19.8\%$$

Step 3: Determine $w(Fe)$ associated with pyrite (Py)

$$w(\text{Fe}) \text{ in Py} = 25.6 - 19.8 = 5.8\%$$

Step 4: Convert Fe to Py (answer to first question)

$$w(\text{Py}) = 5.8\% \left[\frac{55.85 + (2 \times 32.06)}{55.85} \right] = 12.5\%$$

Step 5: Determine w (non-sulfides) (answer to second question)

$$w(\text{non-sulfides}) = 100 - (w(\text{Cp}) + w(\text{Py})) = 100 - (65.0 + 12.5) = 22.5\%$$

1.2 ABUNDANCE OF MINERALS

The price of metals is governed mainly by supply and demand. Supply includes both newly mined and recycled metal, and recycling is now a significant component of the lifecycle of some metals—about 60% of lead supply comes from recycled sources. There have been many prophets of doom over the years pessimistically predicting the imminent exhaustion of mineral supplies, the most extreme perhaps being the "Limits to Growth" report to the Club of Rome in 1972, which forecast that gold would run out in 1981, zinc in 1990, and oil by 1992 (Meadows et al., 1972). Mouat (2011) offers some insights as to the past and future of mining.

In fact, major advances in productivity and technology throughout the twentieth century greatly increased both the resource base and the supply of newly mined metals, through geological discovery and reductions in the cost of production. These advances actually drove down metal prices in real terms, which reduced the profitability of mining companies and had a damaging effect on economies heavily dependent on resource extraction, particularly those in Africa and South America. This in turn drove further improvements in productivity and technology. Clearly mineral resources are finite, but supply and demand will generally balance in such a way that if supplies decline or demand increases, the price will increase, which will motivate the search for new deposits, or technology to render marginal deposits economic, or even substitution by other materials. Gold is an exception, its price having not changed much in real terms since the sixteenth century, due mainly to its use as a monetary instrument and a store of wealth.

Estimates of the crustal abundances of metals are given in Table 1.1 (Taylor, 1964), together with the amounts of some of the most useful metals, to a depth of 3.5km (Tan and Chi-Lung, 1970).

Table 1.1 Abundance of Metal in the Earth's Crust

Element	Abundance/%	Amt. in Top 3.5km/t	Element	Abundance/%	Amt. in Top 3.5km/t
(Oxygen)	46.4		Vanadium	0.014	$10^{14} \sim 10^{15}$
Silicon	28.2		Chromium	0.010	
Aluminum	8.2	$10^{16} \sim 10^{18}$	Nickel	0.0075	
Iron	5.6		Zinc	0.0070	
Calcium	4.1		Copper	0.0055	$10^{13} \sim 10^{14}$
Sodium	2.4		Cobalt	0.0025	
Magnesium	2.3	$10^{16} \sim 10^{18}$	Lead	0.0013	

Continued Table 1.1

Element	Abundance/%	Amt. in Top 3.5km/t	Element	Abundance/%	Amt. in Top 3.5km/t
Potassium	2.1		Uranium	0.00027	
Titanium	0.57	$10^{15} \sim 10^{16}$	Tin	0.00020	
Manganese	0.095		Tungsten	0.00015	$10^{11} \sim 10^{13}$
Barium	0.043		Mercury	8×10^{-6}	
Strontium	0.038		Silver	7×10^{-6}	
Rare earths	0.023		Gold	$<5 \times 10^{-6}$	
Zirconium	0.017	$10^{14} \sim 10^{16}$	Platinum metals	$<5 \times 10^{-6}$	$<10^{11}$

The abundance of metals in the oceans is related to some extent to the crustal abundances, since they have come from the weathering of the crustal rocks, but superimposed upon this are the effects of acid rainwaters on mineral leaching processes; thus, the metal availability from seawater shown in Table 1.2 (Tan and Chi-Lung, 1970) does not follow precisely that of the crustal abundance. The seabed may become a viable source of minerals in the future. Manganese nodules have been known since the beginning of the nineteenth century (Mukherjee et al., 2004), and mineral-rich hydro-thermal vents have been discovered (Scott, 2001). Mining will eventually extend to space as well.

It can be seen from Table 1.1 that eight elements account for over 99% of the earth's crust: 74.6% is silicon and oxygen, and only three of the industrially important metals (aluminum, iron, and magnesium) are present in amounts above 2%. All the other useful metals occur in amounts below 0.1%; copper, for example, which is the most important non-ferrous metal, occurring only to the extent of 0.0055%. It is interesting to note that the so-called common metals, zinc and lead, are less plentiful than the rare-earth metals (cerium, thorium, etc.).

Table 1.2 Abundance of Metal in the Oceans

Element	Abundance/t	Element	Abundance/t
Magnesium	$10^{15} \sim 10^{16}$	Vanadium	$10^{9} \sim 10^{10}$
Silicon	$10^{12} \sim 10^{13}$	Titanium	
Aluminium		Cobalt	
Iron	$10^{10} \sim 10^{11}$	Silver	$10^{12} \sim 10^{13}$
Molybdenum		Tungsten	
Zinc			
Tin		Chromium	
Uranium	$10^{9} \sim 10^{10}$	Gold	$<10^{8}$
Copper		Zirconium	
Nickel		Platinum	

1.3 DEPOSITS AND ORES

It is immediately apparent that if the minerals containing important metals were uniformly distributed throughout the earth, they would be so thinly dispersed that their economic extraction would be impossible. However, the occurrence of minerals in nature is regulated by the geological conditions throughout the life of the mineral. A particular mineral may be found mainly in association with one rock type (for example, cassiterite mainly associates with granite rocks) or may be found associated with both igneous and sedimentary rocks (i. e., those produced by the deposition of material arising from the mechanical and chemical weathering of earlier rocks by water, ice, and chemical decay). Thus, when granite is weathered, cassiterite may be transported and redeposited as an alluvial deposit. Besides these surface processes, mineral deposits are also created due to magmatic, hydrothermal, and other geological events (Ridley, 2013).

Due to the action of these many natural agencies, mineral deposits are frequently found in sufficient concentrations to enable the metals to be profitably recovered; that is, the deposit becomes an ore. Most ores are mixtures of extractable minerals and extraneous nonvaluable material described as gangue. They are frequently classed according to the nature of the valuable mineral. Thus, in native ores the metal is present in the elementary form; sulfide ores contain the metal as sulfides, and in oxidized ores the valuable mineral may be present as oxide, sulfate, silicate, carbonate, or some hydrated form of these. Complex ores are those containing profitable amounts of more than one valuable mineral. Metallic minerals are often found in certain associations within which they may occur as mixtures of a wide range of grain sizes or as single-phase solid solutions or compounds. Galena and sphalerite, for example, are commonly associated, as, to a lesser extent, are copper sulfide minerals and sphalerite. Pyrite is almost always associated with these minerals as a sulfide gangue.

There are several classifications of a deposit, which from an investment point of view it is important to understand: mineral resources are potentially valuable and are further classified in order of increasing confidence into inferred, indicated, and measured resources; mineral (ore) reserves are known to be economically (and legally) feasible for extraction and are further classified, in order of increasing confidence, into probable and proved reserves.

1.4 METALLIC AND NONMETALLIC ORES

Ores of economic value can be classed as metallic or nonmetallic, according to the use of the mineral. Certain minerals may be mined and processed for more than one purpose. In one category, the mineral may be a metal ore, that is, when it is used to prepare the metal, as when bauxite (hydrated aluminum oxide) is used to make aluminum. The alternative is for the compound to be classified as a nonmetallic ore, that is, when bauxite or natural aluminum oxide is used to make material for refractory bricks or abrasives.

Many nonmetallic ore minerals associate with metallic ore minerals (Appendixes I and II) and are mined and processed together. For example, galena, the main source of lead, sometimes associates with fluorite (CaF_2) and barytes ($BaSO_4$), both important nonmetallic minerals.

Diamond ores have the lowest grade of all mined ores. One of the richest mines in terms of diamond content, Argyle (in Western Australia) enjoyed grades as high as 2×10^{-6} in its early life. The lowest grade deposits mined in Africa have been as low as 0.01×10^{-6}. Diamond deposits are mined mainly for their gem quality stones which have the highest value, with the low-value industrial quality stones being essentially a by-product: most industrial diamond is now produced synthetically.

1.5 THE NEED FOR MINERAL PROCESSING

"As-mined" or "run-of-mine" ore consists of valuable minerals and gangue. Mineral processing, also known as ore dressing, ore beneficiation, mineral dressing, or milling, follows mining and prepares the ore for extraction of the valuable metal in the case of metallic ores, or to produce a commercial end product as in the case of minerals such as potash (soluble salts of potassium) and coal. Mineral processing comprises two principal steps: size reduction to liberate the grains of valuable mineral (or paymineral) from gangue minerals, and physical separation of the particles of valuable minerals from the gangue, to produce an enriched portion, or concentrate, containing most of the valuable minerals, and a discard, or tailing (tailings or tails), containing predominantly the gangue minerals. The importance of mineral processing is today taken for granted, but it is interesting to reflect that little more than a century ago, ore concentration was often a fairly crude operation, involving relatively simple density-based and hand-sorting techniques. The twentieth century saw the development of mineral processing as an important profession in its own right, and certainly without it the concentration of many ores, and particularly the metalliferous ores, would be hopelessly uneconomic (Wills and Atkinson, 1991).

It has been predicted that the importance of mineral processing of metallic ores may decline as the physical processes utilized are replaced by the hydro-and pyrometallurgical routes used by the extractive metallurgist (Gilchrist, 1989), because higher recoveries are obtained by some chemical methods. This may apply when the useful mineral is very finely disseminated in the ore and adequate liberation from the gangue is not possible, in which case a combination of chemical and mineral processing techniques may be advantageous, as is the case with some highly complex deposits of copper, lead, zinc, and precious metals (Gray, 1984; Barbery, 1986). Heap leaching of gold and oxidized copper ores are examples where mineral processing is largely by-passed, providing only size reduction to expose the minerals. In-situ leaching is used increasingly for the recovery of uranium and bitumen from their ores. An exciting possibility is using plants to concentrate metals sufficiently for chemical extraction. Known as phytomining or agro-mining, it has shown particular promise for nickel (Moskvitch, 2014). For most ores, however, concentration of metals for subsequent extraction is best accomplished by mineral processing methods that are inexpensive, and their use is readily justified on economic grounds.

The two fundamental operations in mineral processing are, therefore, liberation or release of the valuable minerals from the gangue, and concentration, the separation of these values from the gangue.

1.6 LIBERATION

Liberation of the valuable minerals from the gangue is accomplished by size reduction or comminution, which involves crushing and grinding to such a size that the product is a mixture of relatively clean particles of mineral and gangue, that is, the ore minerals are liberated or free. An objective of comminution is liberation at the coarsest possible particle size. If such an aim is achieved, then not only is energy saved but also by reducing the amount of fines produced any subsequent separation stages become easier and cheaper to operate. If high-grade solid products are required, then good liberation is essential; however, for subsequent hydrometallurgical processes, such as leaching, it may only be necessary to expose the required mineral.

Grinding is often the greatest energy consumer, accounting for up to 50% of a concentrator's energy consumption (Radziszewski, 2013). As it is this process which achieves liberation of values from gangue, it is also the process that is essential for efficient separation of the minerals. In order to produce clean concentrates with little contamination with gangue minerals, it is often necessary to grind the ore to a fine size ($<100\mu m$). Fine grinding increases energy costs and can lead to the production of very fine difficult to treat "slime" particles which may be lost into the tailings, or even discarded before the concentration process. Grinding therefore becomes a compromise between producing clean (high-grade) concentrates, operating costs, and losses of fine minerals. If the ore is low grade, and the minerals have very small grain size and are disseminated through the rock, then grinding energy costs and fines losses can be high.

In practice, complete liberation is seldom achieved, even if the ore is ground down to less than the grain size of the desired minerals. This is illustrated by Fig. 1.1, which shows a lump of ore containing a grain of valuable mineral with a breakage pattern superimposed that divides the lump into cubic particles of identical volume (for simplicity) and of a size below that of the mineral grain. It can be judged that each

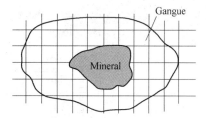

Fig. 1.1 Mineral locked with gangue and illustrating effect of breakage on liberation

particle produced containing mineral also contains a portion of gangue. Complete liberation has not been attained, but the bulk of the major mineral—the gangue—has, however, been liberated from the valuable mineral.

The particles of "locked" (or "composite") mineral and gangue are known as middlings, and further liberation from this fraction can only be achieved by further comminution. The "degree of liberation" refers to the percentage of the mineral occurring as free particles in the broken ore in relation to the total mineral content in locked and free particles. Liberation can be high if there are weak boundaries between mineral and gangue particles, which is often the case with ores composed mainly of rockforming minerals, particularly sedimentary minerals. This is sometimes referred to as "liberation by detachment". Usually, however, the adhesion between mineral and gangue is strong and during comminution the various constituents are cleft across the grain boundaries; that is, breakage is random. Random breakage produces a significant amount of middlings. Approaches to increasing the degree of liberation involve directing the breaking stresses at the mineral grain boundaries, so that the rock can be broken without breaking the mineral grains (Wills and Atkinson, 1993). For example, microwaves can be used, which cause differential heating among the constituent minerals and thus create stress fractures at grain boundaries (Kingman et al., 2004).

Many researchers have tried to quantify (model) the degree of liberation (Barbery, 1991; King, 2012). These models, however, suffer from many unrealistic assumptions that must be made with respect to the grain structure of the minerals in the ore. For this reason liberation models have not found much practical application. However, some fresh approaches by Gay (2004a, b) have demonstrated that there may yet be a useful role for such models.

Fig. 1.2 shows predictions using the simple liberation model based on random breakage derived by Gaudin (1939), but which is sufficient to introduce an important practical point. The degree (fraction) of liberation is given as a function of the particle size to grain size ratio and illustrates that to achieve high liberation, say 75%, the particle size has to be much smaller than the grain size, in this case ca. 1/10th the size. So, for example, if the grain size is 1mm then size reduction must produce a particle size at least 0.1mm (100μm) or less, and if the grain size is 0.1mm the particle size should be 10μm or less. This result helps understand the fine size required from the comminution process. For example, in processing base metal sulfides a target grind size of 100μm for adequate liberation was common in the 1960s but the finer grained ores of today may require a target grind size of 10μm, which in turn has driven the development of new grinding technologies.

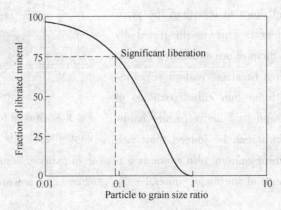

Fig. 1.2 Degree of liberation as a function of the particle to grain ratio

(Derived from model of Gaudin (1939) for the least abundant mineral)

The quantification of liberation is now routine using the dedicated scanning electron microscope systems, for example, and concentrators are increasingly using such systems to monitor the liberation in their processes.

It should be noted that a high degree of liberation is not always necessary and may be undesirable in some cases. For instance, it is possible to achieve a high recovery of values by gravity and magnetic separation even though the valuable minerals are completely enclosed by gangue, and hence the degree of liberation of the values is zero. As long as a pronounced density or magnetic susceptibility difference is apparent between the locked particles and the free gangue particles, the separation is possible. On the other hand, flotation requires at least a surface of the valuable mineral to be exposed.

Fig. 1.3 is a cross section through a typical ore particle and illustrates the liberation dilemma often facing the mineral processor. Regions A represent valuable mineral, and region AA is rich in valuable mineral but is highly intergrown with the gangue mineral. Comminution produces a range of fragments, ranging from fully liberated mineral and gangue particles, to those illustrated. Particles of type 1 are rich in mineral (are high-grade particles) and are classed as concentrate as they have an acceptably low-level degree of locking with the gangue to still make a saleable concentrate grade. Particles of type 4 would be classed as tailings, the small amount of mineral present representing an acceptable loss of mineral. Particles of types 2 and 3 would probably be classed as middlings, although the degree of regrinding needed to promote liberation of mineral from particle 3 would be greater than in particle 2. In practice, ores are ground to an optimum grind size, determined by laboratory and pilot scale testwork, to produce an economic degree of liberation. The concentration process is then designed to produce a concentrate consisting predominantly of valuable mineral, with an accepted degree of locking with the gangue minerals, and a middlings fraction, which may require further grinding to promote optimum release of the minerals. The tailings should be mainly composed of gangue minerals.

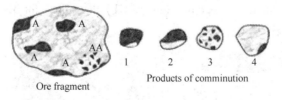

Fig. 1.3 Example cross sections of ore particles

During the grinding of a low-grade ore, the bulk of the gangue minerals is often liberated at a relatively coarse size (see Fig. 1.1). In certain circumstances it may be economic to grind to a size much coarser than the optimum in order to produce, in the subsequent concentration process, a large middlings fraction and tailings which can be discarded at a coarse grain size (Fig. 1.4). The middlings fraction can then be reground to produce a feed to the final concentration process. This method discards most of the coarse gangue early in the process, thus considerably reducing grinding costs.

An intimate knowledge of the mineralogical assembly of the ore ("texture") is essential if efficient processing is to be carried out. Process mineralogy or applied mineralogy thus becomes an important tool for the mineral processor.

Texture refers to the size, dissemination, association, and shape of the minerals within the ore. Processing of the minerals should always be considered in the context of the mineralogy of the ore in order to predict grinding and concentration requirements, feasible concentrate grades, and potential difficulties of separation (Guerney et al., 2003; Baum et al., 2004; Hoal et al., 2009; Evans et al., 2011; Lotter, 2011; Smythe et al., 2013).

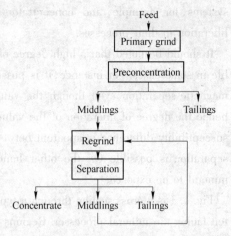

Fig. 1.4 Flowsheet for process utilizing two-stage separation

Microscopic analysis of ores, concentrates, and tailings yields much valuable information regarding the efficiency of the liberation and concentration processes. Fig. 1.5 shows examples of increasing ore complexity, from simple ("free milling" ore, Fig. 1.5(a)) to fine-grained inter-grown texture (Fig. 1.5(c)). Microscopic analysis is particularly useful in troubleshooting problems that arise from inadequate liberation. Conventional optical microscopes can be used for the examination of thin and polished sections of mineral samples, and in mineral sands applications the simple binocular microscope is a practical tool. However, it is now increasingly common to employ quantitative automated mineral analysis using scanning electron microscopy, such as the Mineral Liberation Analyser (MLA) (Gu, 2003), the QEMSCAN (Gottlieb et al., 2000), and the Tescan Integrated Mineral Analyser (TIMA) (Motl et al., 2012), which scan polished sections to give 2D information, and X-ray microcomputed tomography (micro CT) that allows for 3D visualization of particulates (Lin et al., 2013).

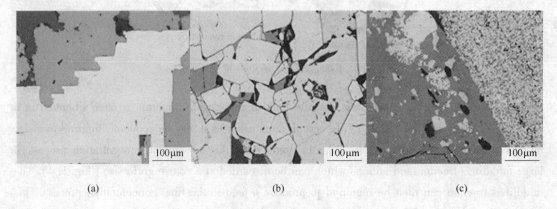

Fig. 1.5 Micrographs showing range in ore texture
(a) galena/sphalerite (Pine Point, Canada); (b) sphalerite/chalcopyrite (Geco, Canada);
(c) galena/sphalerite/pyrite (Mt Isa, Australia) (Courtesy Giovanni Di Prisco)

1.7 CONCENTRATION

After the valuable mineral particles have been liberated, they must be separated from the gangue particles. This is done by exploiting the physical properties of the different minerals. The most important physical properties which are used to concentrate ores are:

(1) Optical. This is often called sorting, which used to be done by hand but is now mostly accomplished by machine (Chapter 7).

(2) Density. Gravity concentration, a technology with its roots in antiquity, is based on the differential movement of mineral particles in water due to their different density and hydraulic properties. The method has seen development of a range of gravity concentrating devices, and the potential to treat dry to reduce reliance on often scarce water resources (Chapter 4). In dense medium separation, particles sink or float in a dense liquid or (more usually) an artificial dense suspension. It is widely used in coal beneficiation, iron ore and diamond processing, and in the preconcentration of some metalliferous ores (Chapter 5).

(3) Surface properties. Froth flotation (or simply "flotation"), which is the most versatile method of concentration, is effected by the attachment of the mineral particles to air bubbles within an agitated pulp. By adjusting the "chemistry" of the pulp by adding various chemical reagents, it is possible to make the valuable minerals water-repellant (hydrophobic) and the gangue minerals water-avid (hydrophilic). This results in separation by transfer of the valuable minerals to the bubbles which rise to form froth on the surface of the pulp.

(4) Magnetic susceptibility. Low-intensity magnetic separators can be used to concentrate strongly magnetic minerals such as magnetite (Fe_3O_4) and pyrrhotite (Fe_7S_8), while high-intensity magnetic separators are used to recover weakly magnetic minerals. Magnetic separation is an important process in the beneficiation of iron ores and finds application in the processing of nonmetallic minerals, such as those found in mineral sand deposits (Chapter 6).

(5) Electrical conductivity. Electrostatic separation can be used to separate conducting minerals from nonconducting minerals. Theoretically this method represents the "universal" concentrating method; virtually all minerals show some difference in conductivity and it should be possible to separate almost any two by this process. However, the method has fairly limited application, and its greatest use is in separating some of the minerals found in heavy sands from beach or stream placers. Minerals must be completely dry and the humidity of the surrounding air must be regulated, since most of the electron movement in dielectrics takes place on the surface and a film of moisture can change the behavior completely. The low capacity of economically sized units is stimulating developments to overcome (Chapter 6).

A general way to show separation is to represent as a recovery to one stream, usually the concentrate, as a function of some mineral property, variously called an efficiency, performance, or partition curve, as illustrated in Fig. 1.6(a). The property can be density, magnetic susceptibility, some measure of hydrophobicity, or particle size (in size separation devices). The plot can be made dimensionless by dividing the property X by X_{50}, the property corresponding to 50% recovery (Fig. 1.6(b)). This is a normalized or reduced efficiency curve. Treating X_{50} as the target property for separation then the ideal or perfect separation is the dashed line in Fig. 1.6(b) passing through $X/X_{50} = 1$.

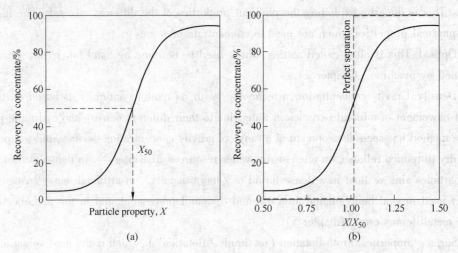

Fig. 1.6 (a) General representation of a physical (mineral) separation process: mineral property can be density, magnetic susceptibility, hydrophobicity, size, etc., and (b) same as (a) but with property made dimensionless

The size of particle is an important consideration in mineral separation. Fig. 1.7 shows the general size range of efficient separation of the concentration processes introduced above. It is evident that all these physical-based techniques fail as the particle size reduces. Extending the particle size range drives innovation.

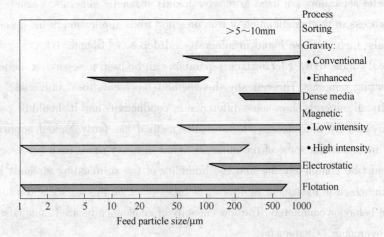

Fig. 1.7 Effective range of application of selected mineral separation techniques
(Adapted from Mills (1978))

In many cases, a combination of two or more separation techniques is necessary to concentrate an ore economically. Gravity separation, for instance, may be used to reject a major portion of the gangue, as it is a relatively cheap process. It may not, however, have the selectivity to produce the final clean concentrate. Gravity concentrates therefore often need further upgrading by more expensive techniques, such as flotation. Magnetic separation can be integrated with flotation—for example, to reject pyrrhotite in processing some Ni-sulfide ores.

Ores which are very difficult to treat (refractory), due to fine dissemination of the minerals, complex mineralogy, or both, have driven technological advances. An example is the zinc-lead-silver deposit at McArthur River, in Australia. Discovered in 1955, it is one of the world's largest zinc-lead deposits, but for 35 years it resisted attempts to find an economic processing route due to the very finegrained texture of the ore. However, the development of the proprietary IsaMill fine grinding technology (Pease, 2005) by the mine's owners Mount Isa Mines, together with an appropriate flotation circuit, allowed the ore to be processed and the mine was finally opened in 1995. The concentrator makes a bulk (i.e., combined) zinc-lead concentrate with a very fine product size of 80% (by weight) finer than $7\mu m$.

Chemical methods can be used to alter mineralogy, allowing the low-cost mineral processing methods to be applied to refractory ores (Iwasaki and Prasad, 1989). For instance, nonmagnetic iron oxides can be roasted in a weakly reducing atmosphere to produce magnetite. In Vale's matte separation process mineral processing (comminution and flotation) is used to separate Ni-Cu matte into separate Cu- and Ni-concentrates which are sent for metal extraction (Damjanovic and Goode, 2000).

Some refractory copper ores containing sulfide and oxidized minerals have been pretreated hydrometallurgically to enhance flotation performance. In the Leach-Precipitation-Flotation process, developed in the years 1929~1934 by the Miami Copper Co., USA, the oxidized minerals are dissolved in sulfuric acid, after which the copper in solution is precipitated as cement copper by the addition of metallic iron. The cement copper and acid-insoluble sulfide minerals are then recovered by flotation. This process, with several variations, has been used at a number of American copper concentrators. A more widely used method of enhancing the flotation performance of oxidized ores is to allow the surface to react with sodium sulfide. This "sulfidization" process modifies the flotation response of the mineral causing it to behave, in effect, as a pseudo-sulfide.

Developments in biotechnology are being exploited in hydrometallurgical operations, particularly in the bacterial oxidation of sulfide gold ores and concentrates (Brierley and Brierley, 2001; Hansford and Vargas, 2001). There is evidence to suggest that certain microorganisms could be used to enhance the performance of conventional mineral processing techniques (Smith et al., 1991). It has been established that some bacteria will act as pyrite depressants in coal flotation, and work has shown that certain organisms can aid flotation in other ways (e.g., Botero et al., 2008). Microorganisms have the potential to profoundly change future industrial flotation practice.

Extremely fine mineral dissemination leads to high energy costs in comminution and losses to tailings due to the generation of difficult-to-treat fine particles. Much research has been directed at minimizing fine mineral losses, either by developing methods of enhancing mineral liberation, thus minimizing the amount of comminution needed, or by increasing the efficiency of conventional physical separation processes, by the use of innovative machines or by optimizing the performance of existing ones. Several methods have been proposed to increase the apparent size of fine particles, by causing them to come together and aggregate. Selective flocculation of certain minerals in suspension, followed by separation of the aggregates from the dispersion, has been achieved on a variety of ore-types at laboratory scale, but plant application is limited.

1.8 REPRESENTING MINERAL PROCESSING SYSTEMS: THE FLOWSHEET

The flowsheet shows diagrammatically the sequence of operations in the plant. In its simplest form it can be presented as a block diagram in which all operations of similar character are grouped (Fig. 1.8). In this case,"comminution" deals with all crushing and grinding. The next block, "separation", groups the various treatments incident to production of concentrate and tailing. The third,"product handling", covers the shipment of concentrates and disposal of tailings.

Expanding, a simple line flowsheet (Fig. 1.9) can be sufficient and can include details of machines, settings, rates, etc. Most flowsheets today use symbols to represent the unit operations. Example flowsheets are given in many of the chapters, with varying degrees of sophistication.

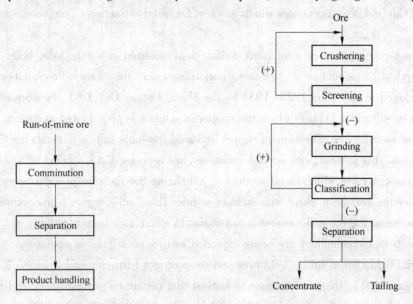

Fig. 1.8 Simple block flowsheet   Fig. 1.9 Line flowsheet

(+) indicates oversized material returned for further treatment and (−) undersized material, which is allowed to proceed to the next stage

1.9 MEASURES OF SEPARATION

The object of mineral processing, regardless of the methods used, is always the same: to separate the minerals with the values in the concentrates, and the gangue in the tailings. The two most common measures of the separation are grade and recovery.

1.9.1　Grade

The grade, or assay, refers to the content of the marketable commodity in any stream, such as the feed and concentrate. In metallic ores, the percent metal is often quoted, although in the case of very low-grade ores, such as gold, metal content may be expressed as parts per million ($\times 10^{-6}$), or its equivalent grams per ton (g/t). Some metals are sold in oxide form, and hence the grade may be quoted in terms of the marketable oxide content: for example, $w(WO_3)$, $w(U_3O_8)$, etc. In nonmetallic operations, grade usually refers to the mineral content: for example, $w(CaF_2)$ in fluorite ores. Diamond ores are usually graded in carats per 100t, where 1 carat is 0.2g. Coal is graded according to its ash content, that is, the amount of incombustible mineral present within the coal. Most coal burned in power stations ("steaming coal") has an ash content between 15% and 20%, whereas "coking coal" used in steel making generally has an ash content of less than 10%.

The metal content of the mineral determines the maximum grade of concentrate that can be produced. Thus processing an ore containing Cu in only chalcopyrite ($CuFeS_2$) the maximum attainable grade is 34.6%, while processing an ore containing galena (PbS), the maximum Pb grade is 86.6% (The method of calculation was explained in Example 1.1).

1.9.2　Recovery

The recovery, in the case of a metallic ore, is the percentage of the total metal contained in the ore that is recovered to the concentrate. For instance, a recovery of 90% means that 90% of the metal in the ore (feed) is recovered in the concentrate and 10% is lost in ("rejected" to) the tailings. The recovery, when dealing with nonmetallic ores, refers to the percentage of the total mineral contained in the ore that is recovered into the concentrate. In terms of the usual symbols recovery R is given by:

$$R = \frac{C \cdot c}{F \cdot f} \tag{1.1}$$

where C is weight of concentrate (or more precisely flowrate, e.g., t/h), F the weight of feed, c the grade (assay) of metal or mineral in the concentrate, and f the grade of metal/mineral in the feed. Provided the metal occurs in only one mineral, metal recovery is the same as the recovery of the associated mineral. (Example 1.2 shows how to deal with situations where an element resides in more than one mineral.) Metal assays are usually given as % but it is often easier to manipulate formulas if the assays are given as fractions (e.g., 10% becomes 0.1).

Related measures to grade and recovery include: the ratio of concentration, the ratio of the weight of the feed (or heads) to the weight of the concentrate (i.e., F/C); weight recovery (or mass or solids recovery) also known as yield is the inverse of the ratio of concentration, that is, the ratio of the weight of concentrate to weight of feed (C/F); enrichment ratio, the ratio of the grade of the concentrate to the grade of the feed (c/f). They are all used as measures of metallurgical efficiency.

1.9.3 Grade-Recovery Relationship

The grade of concentrate and recovery are the most common measures of metallurgical efficiency, and in order to evaluate a given operation it is necessary to know both. For example, it is possible to obtain a very high grade of concentrate (and ratio of concentration) by simply picking a few lumps of pure galena from a lead ore, but the recovery would be very low. On the other hand, a concentrating process might show a recovery of 99% of the metal, but it might also put 60% of the gangue minerals in the concentrate. It is, of course, possible to obtain 100% recovery by not concentrating the ore at all.

Grade of concentrate and recovery are generally inversely related: as recovery increases grade decreases and vice versa. If an attempt is made to attain a very high-grade concentrate, the tailings assays are higher and the recovery is low. If high recovery of metal is aimed for, there will be more gangue in the concentrate and the grade of concentrate and ratio of concentration will decrease. It is impossible to give representative values of recoveries and ratios of concentration. A concentration ratio of 2 to 1 might be satisfactory for certain high-grade nonmetallic ores, but a ratio of 50 to 1 might be considered too low for a low-grade copper ore, and ratios of concentration of several million to one are common with diamond ores. The aim of milling operations is to maintain the values of ratio of concentration and recovery as high as possible, all factors being considered.

Ultimately the separation is limited by the composition of the particles being separated and this underpins the inverse relationship. To illustrate the impact of particle composition, consider the six particle array in Fig. 1.10 is separated in an ideal separator, that is, a separator that recovers particles sequentially based on valuable mineral content. The six particles represent three mineral (A) equivalents and three gangue (B) equivalents (all of equal weight), thus the perfect separator would recover the particles in order 1 through 6. The grade of A and the recovery of A after each particle is separated into the concentrate is given in Table 1.3 and plotted in Fig. 1.11 (which also shows the conversion from mineral grade to metal grade assuming the valuable mineral in this case is chalcopyrite, $CuFeS_2$). The inverse relationship is a consequence of the distribution of particle composition. The figure includes reference to other useful features of the grade recovery relationship: the grade of all mineral A-containing particles, the pure mineral grade approached as recovery goes to zero, and the feed grade, which corresponds to recovery of all particles.

The grade-recovery corresponding to perfect separation is known as the mineralogically-limited or liberation-limited grade-recovery. For simple two component mixtures with well differentiated density, it is possible to approach perfect separation in the laboratory using a sequence of "heavy" liquids (e.g., some organics or concentrated salt solutions) of increasing density (Chapter 5). The liberation-limited curve can be generated from mineralogical data, essentially following the calculations used to generate Fig. 1.11. The liberation-limited grade-recovery curve is used to compare

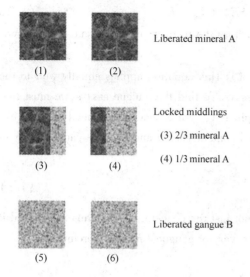

Liberated mineral A

Locked middlings
(3) 2/3 mineral A
(4) 1/3 mineral A

Liberated gangue B

Fig. 1.10 Example particle assemblage fed to perfect separator

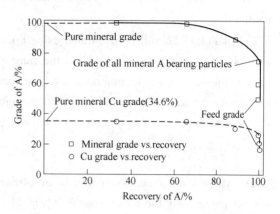

Fig. 1.11 Grade-recovery curve for perfect separation of the six-particle assemblage in Fig. 1.10 showing the inverse relationship (Cu grades based on chalcopyrite, $CuFeS_2$)

Table 1.3 Recovery of Particles in Fig. 1.10 by Perfect Separator

Particle recovered	1	2	3	4	5	6
Concentrate grade/%	100(1/1)①	100 (2/2)	89 (2.66/3)	75 (3/4)	60 (3/5)	50 (3/6)
Recovery/%	33 (1/3)	66 (2/3)	89 (2.66/3)	100 (3/3)	100 (3/3)	100 (3/3)

① Figure in parenthesis is the ratio of the mass of particles.

against the actual operation and determine how far away it is from the theoretical maximum separation.

The grade-recovery curve is the most common representation of metallurgical performance. When assessing the effect of a process change, the resulting grade-recovery curves should be compared, not individual points, and differences between curves should be subjected to tests of statistical significance. To construct the curve, the data need to be collected in increments, such as increments of density or increments of flotation time, and then assembled into cumulative recovery and cumulative grade. There are other methods of presenting the data, such as recovery versus total solids recovery, and recovery of mineral A versus recovery of mineral B. Together these are sometimes referred to as "separability curves" and examples will be found in various chapters.

1.9.4 A Measure of Technical Separation Efficiency

There have been many attempts to combine recovery and concentrate grade into a single index defining the metallurgical efficiency of the separation. These have been reviewed by Schulz (1970), who proposed the following definition:

$$\text{Separation efficiency}(SE) = R_m - R_g \tag{1.2}$$

where R_m is the recovery of the valuable mineral and R_g is the recovery of the gangue into the concentrate.

Calculation of SE will be illustrated using Eq. (1.1). This equation applies equally well to the recovery of gangue, provided we know the gangue assays. To find the gangue assays, we must first convert the metal assay to mineral assay (see Example 1.1). For a concentrate metal grade of c if the metal content in the mineral is m then the mineral content is c/m and thus the gangue assay of the concentrate is given by (assays in fractions):

$$g = 1 - \frac{c}{m} \tag{1.3}$$

This calculation, of course, applies to any stream, not just the concentrate, and thus in the feed if f is the metal assay the mineral assay is f/m. The recovery of gangue (R_g) is therefore

$$R_g = \frac{C}{F} \frac{1 - \dfrac{c}{m}}{1 - \dfrac{f}{m}} \rightarrow \frac{C(m-c)}{F(m-f)} \tag{1.4}$$

and thus the separation efficiency is:

$$SE = \frac{C_c}{F_f} - \frac{C(m-c)}{F(m-f)} \rightarrow \frac{C}{F}\left(\frac{c}{f} - \frac{m-c}{m-f}\right) = \frac{C}{F}\frac{m}{f}\frac{c-f}{m-f} \tag{1.5}$$

Example 1.3 illustrates the calculations. The concept can be extended to consider separation between any pair of minerals, A and B, such as in the separation of lead from zinc.

The concept of separation efficiency is examined in Appendix III, based on Jowett (1975) and Jowett and Sutherland (1985), where it is shown that defining separation efficiency as separation achieved relative to perfect separation yields Eq. (1.5).

Although separation efficiency can be useful in comparing the performance of different operating conditions, it takes no account of economic factors, and is sometimes referred to as the "technical separation efficiency". As will become apparent, a high value of separation efficiency does not necessarily lead to the most economic return. Nevertheless it remains a widely used measure to differentiate alternatives prior to economic assessment.

Example 1.3

A tin concentrator treats a feed containing 1% tin, and three possible combinations of concentrate grade and recovery are:

High grade 63% tin at 62% recovery

Medium grade 42% tin at 72% recovery

Low grade 21% tin at 78% recovery

Determine which of these combinations of grade and recovery produce the highest separation efficiency.

Solution

Assuming that the tin is totally contained in the mineral cassiterite (SnO_2) then $m = 78.6\%$ (0.786), and we can complete the calculation:

Case	$\dfrac{C}{F}$ (from Eq. (1.1))	SE (Eq. (1.5))
High grade	$0.62 = \dfrac{C}{F}\left(\dfrac{0.63}{0.01}\right)$ $\dfrac{C}{F} = 9.841E^{-3}$	$SE = 9.841E^{-3}\left(\dfrac{0.63}{0.01} - \dfrac{0.786 - 0.63}{0.786 - 0.01}\right)$ $= 61.8\%$
Medium grade	$0.72 = \dfrac{C}{F}\left(\dfrac{0.42}{0.01}\right)$ $\dfrac{C}{F} = 1.714E^{-2}$	$SE = 1.714E^{-2}\left(\dfrac{0.42}{0.01} - \dfrac{0.786 - 0.42}{0.786 - 0.01}\right)$ $= 71.2\%$
Low grade	$0.78 = \dfrac{C}{F}\left(\dfrac{0.21}{0.01}\right)$ $\dfrac{C}{F} = 3.714E^{-2}$	$SE = 3.714E^{-2}\left(\dfrac{0.21}{0.01} - \dfrac{0.786 - 0.21}{0.786 - 0.01}\right)$ $= 75.2\%$

The answer to which gives the highest separation efficiency, therefore, is the low-grade case.

1.10 ECONOMIC CONSIDERATIONS

Economic considerations play a large role in mineral processing. The enormous growth of industrialization from the eighteenth century onward led to dramatic increases in the annual output of most mineral commodities, particularly metals. Copper output grew by a factor of 27 in the twentieth century alone, and aluminum by an astonishing factor of 3800 in the same period. Fig. 1.12 shows the world production of aluminum, copper, and zinc for the period 1900~2012 (USGS, 2014).

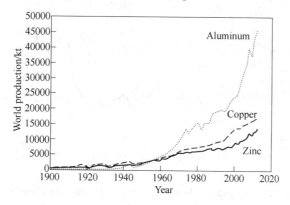

Fig. 1.12 World production of aluminum, copper, and zinc for the period 1900~2012

All of these metals suffered to a greater or lesser extent when the Organization of Petroleum Exporting Countries (OPEC) quadrupled the price of oil in 1973~1974, ending the great postwar industrial boom. The situation worsened in 1979~1981, when the Iranian revolution and then the Iran-Iraq war forced the price of oil up from $13 to nearly $40 a barrel, plunging the world into another and deeper recession. While in the mid-1980s a glut in the world's oil supply as North Sea oil production grew cut the price to below $15 in 1986, Iraq's invasion of Kuwait in 1990 pushed the price up again, to a peak of $42 in October, although by then 20% of the world's energy was being provided by natural gas.

In 1999, overproduction and the Asian economic crisis depressed oil prices to as low as $10 a barrel from where it climbed steadily to a record $147 a barrel in 2008, driven by demand from the now surging Asian economies, particularly China. Turmoil in oil producing regions continued in the twenty-first century from the invasion of Iraq (2003) to the "Arab Spring" (start date 2009) and the festering stand-off with Russia over the Ukraine (2014). Over the past three years, the oil price fluctuated around $100 a barrel, apparently inured against this turmoil. In the last half of 2014, however, the price dropped precipitously to less than $50 a barrel as the world moved to a surplus of oil driven by decreased growth rate in countries like China, and a remarkable, and unexpected, increase in shale oil production in the United States. The decision by OPEC not to reduce production has accelerated the price decline. This will drive out the high-cost producers, including production from shale oil that needs about $60 a barrel to be profitable, and the price will climb.

These fluctuations in oil prices impact mining, due to their influence both on the world economy and thus the demand for metals, and directly on the energy costs of mining and processing. Metal and mineral commodities are thus likewise subject to cycles in price. Fig. 1.13 shows the commodity price index and identifies the recent "super-cycle" starting about 2003, and the decline following that of oil during 2014. These "boom and bust" cycles are characterized by overoptimistic forecasts as prices rise (reference to the "super-cycle") and dire warnings about exporting countries suffering from the "natural resource disease" on the price downslide.

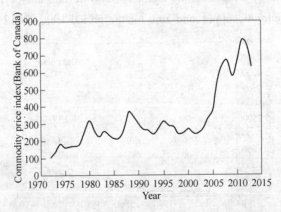

Fig. 1.13 Commodity price index (Bank of Canada) representing the cost of nine commodities
(potash, aluminum, gold, nickel, iron, copper, silver, zinc, and lead)

The cycles spur innovation. To reduce dependence on OPEC, oil production turned to "nonconventional" sources, such as the oil sands in Canada. A radical nonconventional resource, methane chlorates, appeared set to make Siberia the next Saudi Arabia and although interest has waned recently these deposits remain part of our energy future.

The most dramatic new nonconventional source, however, comes from the shale oil deposits in the United States. The vast volumes of natural gas in shale formations in the United States were well known by the 1990s and attempts to crack open the shale in vertical wells by injecting sand, water, and chemicals, known as hydraulic fracturing or "fracking", were underway but had proved unprofitable. The breakthrough innovation was to drill horizontally rather than vertically and thus expose thousands of meters of gas-bearing shale rather than just a few tens of meters (Hefner, 2014). The same technology also released reservoirs of oil in shale and other "tight", that is, impermeable, rock formations. As a consequence, the United States could become the world's largest oil producer by the end of the decade, a revolution with impacts both economic and geopolitical. While other large such shale formations are known throughout the world, environmental concerns over groundwater contamination has, for the moment, cooled the spread of the technology.

The cycles stimulate substitution. In the case of oil, electricity generation turned to natural gas (including liquefied natural gas, LNG) and reconsidered nuclear, while sustaining a large role for coal which remains in plentiful supply and is more generously distributed among the advanced economies. Solar-, wind-, and tidal-generated sources are starting to contribute to the electric power grid. Subsidies, carbon taxing, and cap-and-trade (emissions trading) initiatives spur these "green" alternatives. For transport, we are witnessing a resurgence of interest in electric-powered vehicles. Another energy innovation is "cogeneration", the capture and distribution of waste heat to heat buildings. Mining operations are using this concept, for example, by tapping the natural heating underground and examining the potential to recover the heat generated in comminution (Radziszewski, 2013). The demand to limit greenhouse gas emissions to combat climate change will drive substitution of carbon-based energy sources.

Certain metals and minerals likewise face competition from substitutes and some face additional factors. Avoiding "conflict minerals" influences the choice of supply source. Some companies make an advert of not using commodities seen as environmentally harmful. Lifecycle analysis may influence which commodities are used and in what combinations in order to facilitate recycling.

While commodity prices have gone up in current dollars since 2000 (Fig. 1.13), the capital cost of mining projects has increased by 200%~300% (Thomas et al., 2014) over the same period. This is due to increased equipment and construction costs (competition for materials and labor), increased environmental regulations, and the added complexity of constructing mine sites in remote areas. While the recent high metal prices attracted interest in previously uneconomic deposits, the project cost escalation has put pressure on the mines to optimize performance and has caused investors to be wary.

1.10.1 Contained Value

Every ton of material in the deposit has a certain contained value that is dependent on the metal content and current price of the contained metal. For instance, at a copper price of £2000/t and a molybdenum price of £18/kg, a deposit containing 1% copper and 0.015% molybdenum has a contained value of more than £22/t. To be processed economically, the cost of processing an ore must be less than the contained value.

1.10.2 Processing Costs

Mining is a major cost, and this can vary from only a few pence per ton of ore to well over £50/t.

High-tonnage operations are cheaper in terms of operating costs but have higher initial capital costs. These capital costs are paid off over a number of years, so that high-tonnage operations can only be justified for the treatment of deposits large enough to allow this. Small ore bodies are worked on a smaller scale to reduce overall capital costs, but capital and operating costs per ton are correspondingly higher (Ottley, 1991).

Alluvial mining is the cheapest method and, if on a large scale, can be used to mine ores of low contained value due to low grade or low metal price, or both. For instance, in S. E. Asia, tin ores containing as little as 0.01% Sn are mined by alluvial methods. These ores have a contained value of less than £1/t, but very low processing costs allow them to be economically worked.

High-tonnage open-pit and underground block-caving methods are used to treat ores of low contained value, such as low-grade copper ores. Where the ore must be mined selectively, however, as is the case with underground vein-type deposits, mining methods become very expensive, and can only be justified on ores of high contained value. An underground selective mining cost of £30/t would obviously be hopelessly uneconomic on a tin ore of alluvial grade, but may be economic on a hardrock ore containing 1.5% tin, with a contained value of around £50/t ore.

In order to produce metals, the ore minerals must be broken down by the action of heat (pyrometallurgy), solvents (hydrometallurgy), or electricity (electrometallurgy), either alone or in combination. The most common method is the pyrometallurgical process of smelting. These chemical methods consume large quantities of energy. The smelting of 1t of copper ore, for instance, consumes in the region of 1500 ~ 2000kW·h of electrical energy, which at a cost of say 5 per kW/h is around £85/t, well above the contained value of most copper ores. In addition, smelters are often remote from the mine site, and thus the cost of transport to the site must be considered.

The essential economic purpose of mineral processing is to reduce the bulk of the ore which must be transported to and processed by the smelter, by using relatively cheap, low-energy physical methods to separate the valuable minerals from the gangue minerals. This enrichment process considerably increases the contained value of the ore to allow economic transportation and smelting. Mineral processing is usually carried out at the mine site, the plant being referred to as a mill or concentrator.

Compared with chemical methods, the physical methods used in mineral processing consume relatively small amounts of energy. For instance, to upgrade a copper ore from 1% to 25% metal would use in the region of 20~50kW·h/t. The corresponding reduction in weight of around 25 : 1 proportionally lowers transport costs and reduces smelter energy consumption to around 60 ~ 80kW·h in relation to a ton of mined ore. It is important to realize that, although the physical methods are relatively low energy consumers, the reduction in bulk lowers smelter energy consumption to the order of that used in mineral processing. It is significant that as ore grades decline, the energy used in mineral processing becomes an important factor in deciding whether the deposit is viable to exploit or not.

Mineral processing reduces not only smelter energy costs but also smelter metal losses, due to the production of less metal-bearing slag. (Some smelters include a slag mineral processing stage to recover values and recycle to the smelter) Although technically feasible, the smelting of low-grade ores, apart from being economically unjustifiable, would be very difficult due to the need to produce high-grade metal products free from deleterious element impurities. These impurities are found in the gangue minerals and it is the purpose of mineral processing to reject them into the discard (tailings), as smelters often impose penalties according to their level. For instance, it is necessary to remove arsenopyrite from tin concentrates, as it is difficult to remove the contained arsenic in smelting and the process produces a low-quality tin metal.

Against the economic advantages of mineral processing, the losses occurred during milling and the cost of milling operations must be charged. The latter can vary over a wide range, depending on the method of treatment used, and especially on the scale of the operation. As with mining, large-scale operations have higher capital but lower operating costs (particularly labor and energy) than small-scale operations.

Losses to tailings are one of the most important factors in deciding whether a deposit is viable or not. Losses will depend on the ore mineralogy and dissemination of the minerals, and on the technology available to achieve efficient concentration. Thus, the development of flotation allowed the exploitation of the vast low-grade porphyry copper deposits which were previously uneconomic to treat (Lynch et al. ,2007). Similarly, the introduction of solvent extraction enabled Nchanga Consolidated Copper Mines in Zambia to treat 9Mt per year of flotation tailings, to produce 80000t of finished copper from what was previously regarded as waste (Anon. ,1979).

In many cases not only is it necessary to separate valuable from gangue minerals, but also to separate valuable minerals from each other. For instance, complex sulfide ores containing economic amounts of copper, lead, and zinc usually require separate concentrates of the minerals of each of these metals. The provision of clean concentrates, with little contamination with associated metals, is not always economically feasible, and this leads to another source of loss other than direct tailing loss. A metal which reports to the "wrong" concentrate may be difficult, or economically impossible, to recover and never achieves its potential valuation. Lead, for example, is essentially irrecoverable in copper concentrates and is often penalized as an impurity by the copper smelter. The treatment of such polymetallic base metal ores, therefore, presents one of the great challenges to the mineral processor.

Mineral processing operations are often a compromise between improvements in metallurgical efficiency and milling costs. This is particularly true with ores of low contained value, where low milling costs are essential and cheap unit processes are necessary, particularly in the early stages, where the volume of material treated is high. With such low-value ores, improvements in metallurgical efficiency by the use of more expensive methods or reagents cannot always be justified. Conversely, high metallurgical efficiency is usually of most importance with ores of high contained value and expensive high-efficiency processes can often be justified on these ores.

Apart from processing costs and losses, other costs which must be taken into account are indirect costs such as ancillary services—power supply, water, roads, tailings disposal—which will depend much on the size and location of the deposit, as well as taxes, royalty payments, investment requirements, research and development, medical and safety costs, etc.

1.10.3 Milling Costs

As discussed, the balance between milling costs and metal losses is crucial, particularly with low-grade ores, and because of this most mills keep detailed accounts of operating and maintenance costs, broken down into various subdivisions, such as labor, supplies, energy, etc. for the various areas of the plant. This type of analysis is used to identify high-cost areas where improvements in performance would be most beneficial. It is impossible to give typical operating costs for milling operations, as these vary considerably from mine to mine, and particularly from country to country, depending on local costs of energy, labor, water, supplies, etc.

Table 1.4 is an approximate breakdown of costs for a 100000t/d copper concentrator. Note the dominance of grinding, due mainly to power requirements.

Table 1.4 Relative Costs for a 100000t/d Copper Concentrator

Item	Cost/%
Crushing	2.8
Grinding	47.0
Flotation	16.2
Thickening	3.5
Filtration	2.8
Tailings	5.1
Reagents	0.5
Pipeline	1.4
Water	8.0
Laboratory	1.5
Maintenance support	0.8
Management support	1.6
Administration	0.6
Other expenses	8.1
Total	100

1.10.4 Tailings Reprocessing and Recycling

Mill tailings which still contain valuable components constitute a potential future resource. New or improved technologies can allow the value contained in tailings, which was lost in earlier processing, to be recovered, or commodities considered waste in the past to become valuable in a new economic order. Reducing or eliminating tailings dumps by retreating them also reduces the environmental impact of the waste.

The cost of tailings retreatment is sometimes lower than that of processing the original ore, because much of the expense has already been met, particularly in mining and comminution. There are many tailings retreatment plants in a variety of applications around the world. The East Rand

Gold and Uranium Company closed its operations in 2005 after 28 years having retreated over 870Mt of the iconic gold dumps of Johannesburg, significantly modifying the skyline of the Golden City and producing 250t of gold in the process. Also in 2005, underground mining in Kimberley closed, leaving a tailings dump retreatment operation as the only source of diamond production in the Diamond City. Some platinum producers in South Africa now operate tailings retreatment plants for the recovery of platinum group metals (PGMs), and also chromite as a byproduct from the chrome-rich UG2 Reef. The tailings of the historic Timmins gold mining area of Canada are likewise being reprocessed.

Although these products, particularly gold, tend to dominate the list of tailings retreatment operations because of the value of the product, there are others, both operating and being considered as potential major sources of particular commodities. For example: coal has been recovered from tailings in Australia (Clark, 1997), uranium is recovered from copper tailings by the Uranium Corporation of India, and copper has been recovered from the Bwana Mkubwa tailings in Zambia, using solvent extraction and electrowinning. The Kolwezi Tailings project in the DRC (Democratic Republic of Congo) that recovered oxide copper and cobalt from the tailings of 50 years of copper mining ran from 2004 to 2009. Phytomining, the use of plants to accumulate metals, could be a low-cost way to detoxify tailings (and other sites) and recover metals. Methods of resource recovery from metallurgical wastes are described by Rao (2006).

Recovery from tailings is a form of recycling. The reprocessing of industrial scrap and domestic waste for metal recycling is a growing economic, and environmental, activity. "Urban ore" is a reference to forgotten supplies of metals that lie in and under city streets. Recovery of metals from electronic scrap is one example; another is recovery of PGMs that accumulate in road dust as car catalytic converters wear (Ravilious, 2013). Many mineral separation techniques are applicable to processing urban ores but tapping them is held back by their being so widely spread in the urban environment. The principles of recycling are comprehensively reviewed by Reuter et al. (2005).

1.10.5 Net Smelter Return and Economic Efficiency

Since the purpose of mineral processing is to increase the economic value of the ore, the importance of the grade recovery relationship is in determining the most economic combination of grade and recovery that will produce the greatest financial return per ton of ore treated in the plant. This will depend primarily on the current price of the valuable product, transportation costs to the smelter, refinery, or other further treatment plant, and the cost of such further treatment, the latter being very dependent on the grade of concentrate supplied. A high-grade concentrate will incur lower smelting costs, but the associated lower recovery means lower returns of final product. A low-grade concentrate may achieve greater recovery of the values, but incur greater smelting and transportation costs due to the included gangue minerals. Also of importance are impurities in the concentrate which may be penalized by the smelter, although precious metals may produce a bonus.

The net return from the smelter (NSR) can be calculated for any grade recovery combination from:

NSR = Payment for contained metal − (Smelter charges + Transport costs)   (1.6)

This is summarized in Fig. 1.14, which shows that the highest value of NSR is produced at an optimum concentrate grade. It is essential that the mill achieves a concentrate grade that is as close as possible to this target grade. Although the effect of moving slightly away from the optimum may only be of the order of a few pence per ton treated, this can amount to very large financial losses, particularly on high-capacity plants treating thousands of tons per day. Changes in metal price, smelter terms, etc. obviously affect the NSR versus concentrate grade relationship and the value of the optimum concentrate grade. For instance, if the metal price increases, then the optimum grade will be lower, allowing higher recoveries to be attained (Fig. 1.15).

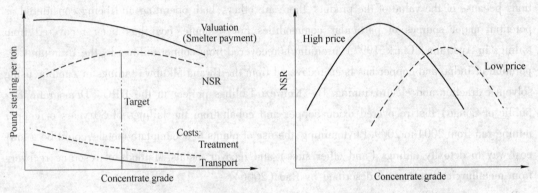

Fig. 1.14 Variation of payment and charges (costs) with concentrate grade

Fig. 1.15 Effect of metal price on NSR grade relationship

It is evident that the terms agreed between the concentrator and the smelter are of paramount importance in the economics of mining and milling operations. Such smelter contracts are usually fairly complex. Concentrates are sold under contract to "custom smelters" at prices based on quotations on metal markets such as the London Metal Exchange (LME). The smelter, having processed the concentrates, disposes of the finished metal to the consumers. The proportion of the "free market" price of the metal received by the mine is determined by the terms of the contract negotiated between mine and smelter, and these terms can vary widely. Table 1.5 summarizes a typical low-grade smelter contract for the purchase of tin concentrates. As is usual in many contracts, one assay unit (1%) is deducted from the concentrate assay in assessing the value of the concentrates, and arsenic present in the concentrate (in this case) is penalized. The concentrate assay is of prime importance in determining the valuation, and the value of the assay is usually agreed on the result of independent sampling and assaying performed by the mine and smelter. The assays are compared, and if the difference is no more than an agreed value, the mean of the two results may be taken as the agreed assay. In the case of a greater difference, an "umpire" sample is assayed at an independent laboratory. This umpire assay may be used as the agreed assay, or the mean of this assay and that of the party which is nearer to the umpire assay may be chosen.

The use of smelter contracts, and the importance of the by-products and changing metal prices, can be seen by briefly examining the economics of processing two base metals-tin and copper-whose fortunes have fluctuated over the years for markedly different reasons.

Table 1.5 Simplified Tin Smelter Contract

Material	Tin concentrates, assaying no less than 15% Sn, to be free from deleterious impurities not stated and to contain sufficient moisture as to evolve no dust when unloaded at our works
Quantity	Total production of concentrates
Valuation	Tin, less 1 unit per dry ton of concentrates, at the lowest of the official LME prices
Pricing	On the seventh market day after completion of arrival of each sampling lot into our works
Treatment charge	£385 per dry ton of concentrates
Moisture	£24/t of moisture
Penalties	Arsenic £40 per unit per ton
Lot charge	£175 per lot sampled of less than 17t
Delivery	Free to our works in regular quantities, loose on a tipping lorry (truck) or in any other manner acceptable to both parties

1.10.6 Case Study: Economics of Tin Processing

Tin constitutes an interesting case study in the vagaries of commodity prices and how they impact the mineral industry and its technologies. Almost half the world's supply of tin in the mid-nineteenth century was mined in southwest England, but by the end of the 1870s Britain's premium position was lost, with the emergence of Malaysia as the leading producer and the discovery of rich deposits in Australia. By the end of the century, only nine mines of any consequence remained in Britain, where 300 had flourished 30 years earlier. From alluvial or secondary deposits, principally from South-East Asia, comes 80% of mined tin. Unlike copper, zinc, and lead, production of tin has not risen dramatically over the years and has rarely exceeded 250000t per annum.

The real price of tin spent most of the first half of the twentieth century in a relatively narrow band between US \$10000 and US \$15000/t (1998 \$), with some excursions (Fig. 1.16). From 1956 its price was regulated by a series of international agreements between producers and consumers under the auspices of the International Tin Council (ITC), which mirrored the highly suc-

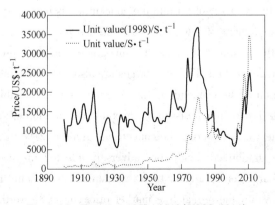

Fig. 1.16 Tin price 1900~2012 (USGS, 2014)

cessful policy of De Beers in controlling the gem diamond trade. Price stability was sought through selling from the ITC's huge stockpiles when the price rose and buying into the stockpile when the price fell.

From the mid-1970s, however, the price of tin was driven artificially higher at a time of world recession, a toxic combination of expanding production and falling consumption, the latter due mainly to the increasing use of aluminum in making cans, rather than tin-plated steel. Although the ITC imposed restrictions on the amount of tin that could be produced by its member countries, the reason for the inflating tin price was that the price of tin was fixed by the Malaysian dollar, while the buffer stock manager's dealings on the LME were financed in sterling. The Malaysian dollar was tied to the American dollar, which strengthened markedly between 1982 and 1984, having the effect of increasing the price of tin in London simply because of the exchange rate. However, the American dollar began to weaken in early 1985, taking the Malaysian dollar with it, and effectively reducing the LME tin price from its historic peak. In October 1985, the buffer stock manager announced that the ITC could no longer finance the purchase of tin to prop up the price, as it had run out of funds, owing millions of pounds to the LME traders. This announcement caused near panic, the tin price fell to £8140/t and the LME halted all further dealings. In 1986, many of the world's tin mines were forced to close due to the depressed tin price, and prices continued to fall in subsequent years, rising again in concert with other metals during the "super-cycle". While the following discussion relates to tin processing prior to the collapse, including prices and costs, the same principles can be applied to producing any metal-bearing mineral commodity at any particular period including the present day.

It is fairly easy to produce concentrates containing over 70% tin (i.e., over 90% cassiterite) from alluvial ores, such as those worked in South-East Asia. Such concentrates present little problem in smelting and hence treatment charges are relatively low. Production of high-grade concentrates also incurs relatively low freight charges, which is important if the smelter is remote. For these reasons it has been traditional in the past for hard-rock, lode tin concentrators to produce high-grade concentrates, but high tin prices and the development of profitable low-grade smelting processes changed the policy of many mines toward the production of lower-grade concentrates. The advantage of this is that the recovery of tin into the concentrate is increased, thus increasing smelter payments. However, the treatment of low-grade concentrates produces much greater problems for the smelter, and hence the treatment charges at "low-grade smelters" are normally much higher than those at the high-grade smelters. Freight charges are also correspondingly higher. Example 1.4 illustrates the identification of the economic optimum grade-recovery combination.

This result in Example 1.4 is in contrast to the maximum separation efficiency which was for the low-grade case (Example 1.3). Lowering the concentrate grade to 21% tin, in order to increase recovery, increased the separation efficiency, but adversely affected the economic return from the smelter, the increased charges being greater than the increase in revenue from the metal.

In terms of contained value, the ore, at free market price, has £85 worth of tin per ton ((1%/100)×£8500/t (of tin)); thus even at the economic optimum combination, the mine realizes only

62% of the ore value in payments received (£52.80/£85).

The situation may alter, however, if the metal price changes appreciably. If the tin price falls and the terms of the smelter contract remain the same, then the mine profits will suffer due to the reduction in payments. Rarely does a smelter share the risks of changing metal price, as it performs a service role, changes in smelter terms being made more on the basis of changing smelter costs rather than metal price. The mine does, however, reap the benefits of increasing metal price.

At a tin price of £6500/t, the NSR per ton of ore from the low-grade smelter treating the 42% tin concentrate is £38.75, while the return from the high-grade smelter, treating a 63% Sn concentrate, is £38.96. Although this is a difference of only £0.21/t of ore, to a small 500t/d tin concentrator this change in policy from relatively low- to high-grade concentrate, together with the subsequent change in concentrate market, would expect to increase the revenue by £0.21×500×365 = £38325 per annum. The concentrator management must always be prepared to change its policies, both metallurgical and marketing, if maximum returns are to be made, although generation of a reliable grade-recovery relationship is often difficult due to the complexity of operation of lode tin concentrators and variations in feed characteristics.

It is, of course, necessary to deduct the costs of mining and processing from the NSR in order to deduce the profit achieved by the mine. Some of these costs will be indirect, such as salaries, administration, research and development, medical and safety, as well as direct costs, such as operating and maintenance, supplies and energy. The breakdown of milling costs varies significantly from mine to mine, depending on the size and complexity of the operations (Table 1.4 is one example breakdown). Mines with large ore reserves tend to have high throughputs, and so although the capital outlay is higher, the operating and labor costs tend to be much lower than those on smaller plants, such as those treating lode tin ores. Mining costs also vary considerably and are much higher for underground than for open-pit operations.

Example 1.4

Suppose that a tin concentrator treats a feed containing 1% tin, and that three possible combinations of concentrate grade and recovery are those in Example 1.3:

High grade 63% tin at 62% recovery

Medium grade 42% tin at 72% recovery

Low grade 21% tin at 78% recovery

Using the low-grade smelter terms set out in the contract in Table 1.5, and assuming that the concentrates are free of arsenic, and that the cost of transportation to the smelter is £20/t of dry concentrate, what is the combination giving the economic optimum, that is, maximum NSR?

Solution

As a common basis we use 1t of ore (feed). For each condition it is possible to calculate the amount of concentrate that can be made from 1t of ore and from this, the smelter payment and charges for each combination of grade and recovery can be calculated.

The calculation of the amount of concentrate follows from the definition of recovery (Eq. (1.1)):

$$R = \frac{Cc}{Ff}; \quad F = 1 \to C = \frac{(1 \times f) \times R}{c}$$

For a tin price of £8500/t (of tin), and a treatment charge of £385/t (of concentrate) the NSR for the three cases is calculated in the table below:

Case	Weight of concentrate/kg · t^{-1}	Smelter payment[①]	Charge		NSR
			Treatment	Transport	
High grade	$\frac{(1t \times 1\%) \times 62\%}{63\%} = 9.84$kg	$\frac{P \times 9.84 \times (63-1)}{100000} = £51.86$	$\frac{9.84 \times 385}{1000} = £3.79$	£0.20	£47.88
Medium grade	$\frac{(1t \times 1\%) \times 72\%}{42\%} = 17.14$kg	$\frac{P \times 17.14 \times (42-1)}{100000} = £59.73$	$\frac{17.14 \times 385}{1000} = £6.59$	£0.34	£52.80
Low grade	$\frac{(1t \times 1\%) \times 78\%}{21\%} = 37.14$kg	$\frac{P \times 37.14 \times (21-1)}{100000} = £63.14$	$\frac{37.14 \times 385}{1000} = £14.30$	£0.74	£48.10

①Units of 10 are to convert from %. The optimum combination is thus the second case.

If mining and milling costs of £40 and £8, respectively, per ton of ore are typical of underground tin operations, then it can be seen that at a tin price of £8500, the mine producing a concentrate of 42% tin, which is sold to a low-grade smelter, makes a profit of £52.80−£48 = £4.80/t of ore. It is also clear that if the tin price falls to £6500/t, the mine loses £48−£38.96 = £9.04 for every = £ ton of ore treated.

The mine profit per ton of ore treated can be illustrated by considering "contained values". For the 72% recovery case (medium-grade case in Example 1.4) the contained value in the concentrate is £85×0.72 = £61.20, and thus the contained value lost in the tailings is $23.80 (£85−£61.20). Since the smelter payment is £52.80 the effective cost of transport and smelting is £61.20−£52.80 = £8.40. Thus the mine profit can be summarized as follows:

Contained value of ore − (costs + losses)

Which for the 72% recovery case is:

$$£[85 - (8.4 + 40 + 23.8)] = £4.80/t$$

The breakdown of revenue and costs in this manner is summarized in Fig. 1.17.

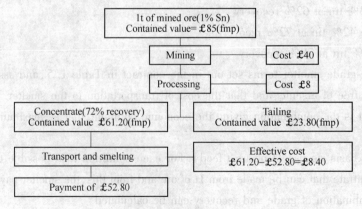

Fig. 1.17 Breakdown of revenues and costs for treatment of lode tin

(fmp = free market price)

In terms of effective cost of production, since 1t of ore produces 0.0072t of tin in concentrates, and the free market value of this contained metal is £61.20 and the profit is £4.80, the total effective cost of producing 1t of tin in concentrates is £(61.20−4.80)/0.0072= £7833.

The importance of metal losses in tailings is shown clearly in Fig. 1.17. With ore of relatively high contained value, the recovery is often more important than the cost of promoting that recovery. Hence relatively high-cost unit processes can be justified if significant improvements in recovery are possible, and efforts to improve recoveries should always be made. For instance, suppose the concentrator, maintaining a concentrate grade of 42% tin, improves the recovery by 1%, i.e., to 73%, with no change in actual operating costs. The NSR will be £53.53/t of ore and after deducting mining and milling costs, the profit realized by the mine will be £5.53/t of ore. Since 1t of ore now produces 0.0073t of tin, having a contained value of £62.05, the cost of producing 1t of tin in concentrates is thereby reduced to £(62.05−5.53)/0.0073=£7742.

Due to the high processing costs and losses, hard-rock tin mines, such as those in Cornwall and Bolivia, had the highest production costs, being above £7500/t of ore in 1985 (for example). Alluvial operations, such as those in Malaysia, Thailand, and Indonesia, have lower production costs (around £6000/t in 1985). Although these ores have much lower contained values (only about £1~2/t), mining and processing costs, particularly on the large dredging operations, are low, as are smelting costs and losses, due to the high concentrate grades and recoveries produced. In 1985, the alluvial mines in Brazil produced the world's cheapest tin, having production costs of only about £2200/t of ore (Anon.,1985a).

1.10.7 Case Study: Economics of Copper Processing

In 1835, the United Kingdom was the world's largest copper producer, mining around 15000t per annum, just below half the world production. This leading position was held until the mid-1860s when the copper mines of Devon and Cornwall became exhausted and the great flood of American copper began to make itself felt. The United States produced about 10000t in 1867, but by 1900 was producing over 250000t per annum. This output had increased to 1000000t per annum in the mid-1950s, by which time Chile and Zambia had also become major producers. World annual production now exceeds 15000000t (Fig. 1.12).

Fig. 1.18 shows that the price of copper in real terms grew steadily from about 1930 until the period of the oil shocks of the mid-1970s and then declined steeply until the early twenty-first century, the average real price in 2002 being lower than that at any time in the twentieth century. The pressure on costs was correspondingly high, and the lower cost operators such as those in Chile had more capacity to survive than the high-cost producers such as some of those in the United States. However, world demand, particularly from emerging economies such as China, drove the price strongly after 2002 and by 2010 it had recovered to about US$9000/t(in dollars of the day) before now sliding back along with oil and other commodities.

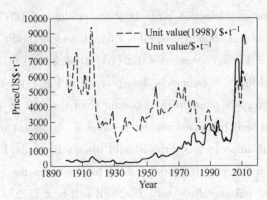

Fig. 1.18 Copper price 1900~2012(USGS, 2014)

The move to large-scale operations (Chile's Minerara Escondida's two concentrators had a total capacity of 230000t/d in 2003), along with improvements in technology and operating efficiencies, have kept the major companies in the copper-producing business. In some cases by-products are important revenue earners. BHP Billiton's Olympic Dam produces gold and silver as well as its main products of copper and uranium, and Rio Tinto's Kennecott Utah Copper is also a significant molybdenum producer.

A typical smelter contract for copper concentrates is summarized in Table 1.6. As in the case of the tin example, the principles illustrated can be applied to current prices and costs.

Table 1.6 Simplified Copper Smelter Contract

Payments	
Copper	Deduct from the agreed copper assay 1 unit, and pay for the remainder at the LME price for higher grade copper
Silver	If over 30g/t pay for the agreed silver content at 90% of the LME silver price
Gold	If over 1g/t pay for the agreed gold content at 95% of the LME gold price
Deductions	
Treatment charge: £30 per dry ton of concentrates	
Refining charge: £115/t of payable copper	

Consider a porphyry copper mine treating an ore containing 0.6% Cu to produce a concentrate containing 25% Cu, at 85% recovery. This represents a concentrate production of 20.4kg/t of ore treated. Therefore, at a copper price of £980/t:

$$\text{Payment for copper} = £980 \times \frac{20.4}{1000} \times 0.24 = £4.80$$

$$\text{Treatment charge} = £30 \times \frac{20.4}{1000} = £0.61$$

$$\text{Refining charge} = £115 \times \frac{20.4}{1000} \times 0.24 = £0.56$$

Assuming a transport cost of £20/t of concentrate, the total deductions are £(0.61+0.56+

0.41) = £1.58, and the NSR per ton of ore treated is thus £(4.80−1.58) = £3.22.

As mining, milling and other costs must be deducted from this figure, it is apparent that only those mines with very low operating costs can have any hope of profiting from such low-grade operations at the quoted copper price. Assuming a typical large open-pit mining cost of £1.25/t of ore, a milling cost of £2/t and indirect costs of £2/t, the mine will lose £2.03 for every ton of ore treated. The breakdown of costs and revenues, following the example for tin (Fig. 1.17), is given in Fig. 1.19.

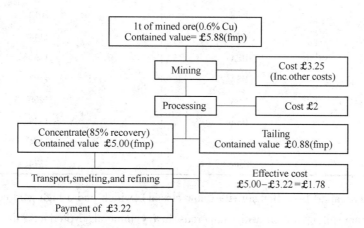

Fig. 1.19 Breakdown of costs and revenues for treatment of typical porphyry copper ore

As each ton of ore produces 0.0051t of copper in concentrates, with a free market value of £5.00, the total effective production costs are £(5.00+2.03)/0.0051t = £1378/t of copper in concentrates. However, if the ore contains appreciable by-products, the effective production costs are reduced. Assuming the concentrate contains 25g/t of gold and 70g/t of silver, then the payment for gold, at a LME price of £230/troy oz (1 troy oz = 31.1035g) is:

$$\frac{20.4}{1000} \times \frac{25}{31.1035} \times 0.95 \times 230 = £3.58$$

and the payment for silver at a LME price of £4.5 per troy oz

$$\frac{20.4}{1000} \times \frac{70}{31.1035} \times 0.95 \times 4.5 = £0.19$$

The net smelter return is thus increased to £6.99/t of ore, and the mine makes a profit of £1.7/t of ore treated. The effective cost of producing 1t of copper is thus reduced to £(5.00−1.74)/0.0051 = £639.22.

By-products are thus extremely important in the economics of copper production, particularly for very low-grade operations. In this example, 42% of the mine's revenue is from gold. This compares with the contributions to revenue realized at Bougainville Copper Ltd. (Sassos, 1983).

Table 1.7 lists estimated effective costs per ton of copper processed in 1985 at some of the world's major copper mines, at a copper price of £980/t (Anon., 1985b).

Table 1.7 Effective Costs at World's Leading Copper Mines in 1985

Mine	Country	Effective Cost of Processed Cu/£ · t^{-1}
Chuquicamata	Chile	589
Bougainville	Papua New Guinea	664
Palabora	South Africa	725
Andina	Chile	755
Cuajone	Peru	876
Toquepala	Peru	1012
Inspiration	USA	1148
San Manuel	USA	1163
Morenci	USA	1193
Twin Buttes	USA	1208
Utah/Bingham	USA	1329
Nchanga	Zambia	1374
Gecamines	Zaire	1374

It is evident that, apart from Bougainville (now closed), which had a high gold content, and Palabora, a large open-pit operation with numerous heavy mineral byproducts, the only economic copper mines in 1985 were the large South American porphyries. Those mines profited due to relatively low actual operating costs, byproduct molybdenum production, and higher average grades (1.2% Cu) than the North American porphyries, which averaged only 0.6% Cu. Relatively high-grade deposits such as at Nchanga failed to profit due partly to high operating costs, but mainly due to the lack of byproducts. It is evident that if a large copper mine is to be brought into production in such an economic climate, then initial capitalization on high-grade secondary ore and byproducts must be made, as at Ok Tedi in Papua New Guinea, which commenced production in 1984, initially mining and processing the high-grade gold ore in the leached capping ore.

Since the profit margin involved in the processing of modern copper ores is usually small, continual efforts must be made to try to reduce milling costs and metal losses. This need has driven the increase in processing rates, many over 100000t ore per day, and the corresponding increase in size of unit operations, especially grinding and flotation equipment. Even relatively small increases in return per ton can have a significant effect, due to the very large tonnages that are often treated. There is, therefore, a constant search for improved flowsheets and flotation reagents.

Fig. 1.19 shows that in the example quoted, the contained value in the flotation tailings is £0.88/t treated ore. The concentrate contains copper to the value of £5.00, but the smelter payment is £3.22. Therefore, the mine realizes only 64.4% of the free market value of copper in the concentrate. On this basis, the actual metal loss into the tailings is only about £0.57/t of ore. This is relatively small compared with milling costs, and an increase in recovery of 0.5% would raise the net smelter return by only £0.01. Nevertheless, this can be significant; to a mine treating 50000t/d, this is an increase in revenue of £500/d, which is extra profit, providing that it is not offset by any increased milling costs. For example, improved recovery in flotation may be possible by the use of a more effective reagent or by increasing the dosage of an existing reagent, but if the increased reagent cost is greater than the increase in smelter return, then the action is not justified.

This balance between milling cost and metallurgical efficiency is critical on a concentrator treating an ore of low contained value, where it is crucial that milling costs be as low as possible. Reagent costs are typically around 10% of the milling costs on a large copper mine, but energy costs may contribute well over 25% of these costs. Grinding is by far the greatest energy consumer and this process undoubtedly has the greatest influence on metallurgical efficiency. Grinding is essential for the liberation of the minerals in the ore, but it should not be carried out any finer than is justified economically. Not only is fine grinding energy intensive, but it also leads to increased media (grinding steel) consumption costs. Grinding steel often contributes as much as, if not more than, the total mill energy cost, and the quality of grinding medium used often warrants special study. Fig. 1. 20 shows the effect of fineness of grind on NSR and grinding costs for a typical low-grade copper ore. Although flotation recovery, and hence NSR, increases with fineness of grind, it is evident that there is no economic benefit in grinding finer than a certain grind size (80% passing size). Even this fineness may be beyond the economic limit because of the additional capital cost of the grinding equipment required to achieve it.

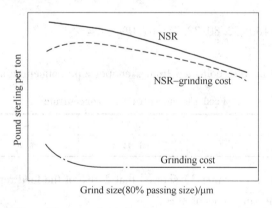

Fig. 1. 20 Effect of fineness of grind on net smelter return and grinding costs

1. 10. 8 Economic Efficiency

It is apparent that a certain combination of grade and recovery produces the highest economic return under certain conditions of metal price, smelter terms, etc. Economic efficiency compares the actual NSR per ton of ore milled with the theoretical return, thus taking into account all the financial implications. The theoretical return is the maximum possible return that could be achieved, assuming "perfect milling", that is, complete separation of the valuable mineral into the concentrate, with all the gangue reporting to tailings. Using economic efficiency, plant efficiencies can be compared even during periods of fluctuating market conditions (Example 1. 5).

Example 1.5

Calculate the economic efficiency of a tin concentrator, treating an ore grading 1% tin producing a concentrate.

Grading 42% tin at 72% recovery, under the conditions of the smelter contract shown in Table 1.5. The cost of transportation to the smelter is £20/t of concentrate. Assume a tin price of £8500/t.

Solution

It was shown in Example 1.4 that this concentrate would realize a net smelter return of £52.80. Assuming perfect milling, 100% recovery of the tin would be achieved into a concentrate grading 78.6% tin (i.e., pure cassiterite), thus the calculations proceed as follows:

The weight of concentrate produced from 1t of feed = 12.72kg

Therefore, transport cost = $£20 \times \dfrac{12.72}{1000} = £0.25$

Treatment charge = $£385 \times \dfrac{12.72}{1000} = £4.90$

Valuation = $£12.72 \times (78.6 - 1) \times \dfrac{8500}{100000} = £83.90$

Therefore, the "perfect milling" net smelter return = £(83.90 - 4.90 - 0.25) = £78.75, and thus:

Economic efficiency = 100 × 52.80/78.75 = 67.0%

Example 1.6

The following assay data were collected from a copper-zinc concentrator:

		Feed	Cu concentrate	Zn concentrate
Assay	Cu/%	0.7	24.6	0.4
	Zn/%	1.94	3.40	49.7

Mass balance calculation (Chapter 3) showed that 2.6% of the feed reported to the copper concentrate and 3.5% to the zinc concentrate.

Calculate the overall economic efficiency under the following simplified smelter terms:

Copper:

Copper price: £1000/t

Smelter payment: 90% of Cu content

Smelter treatment charge: £30/t of concentrate

Transport cost: £20/t of concentrate

Zinc:

Zinc price: £400/t

Smelter payment: 85% of zinc content

Smelter treatment charge: £100/t of concentrate

Transport cost: £20/t of concentrate

Solution

1. Assuming perfect milling

a. Copper

Assuming that all the copper is contained in the mineral chalcopyrite, then maximum copper grade is 34.6% Cu (pure chalcopyrite, $CuFeS_2$).

If C is weight of copper concentrate per 1000kg of feed, then for 100% recovery of copper into this concentrate:

$$100 = \frac{34.6 \times C \times 100}{0.7 \times 1000}, \quad C = 20.2(kg)$$

Transport cost = £20×20.2/1000 = £0.40

Treatment cost = £30×20.2/1000 = £0.61

Revenue = £20.2×0.346×1000×0.9/1000 = £6.29

Therefore, NSR for copper concentrate = £5.28/t of ore

b. Zinc

Assuming that all the zinc is contained in the mineral sphalerite, maximum zinc grade is 67.1% (assuming sphalerite is ZnS).

If Z is weight of zinc concentrate per 1000kg of feed, then for 100% recovery of zinc into this concentrate:

$$100 = \frac{67.1 \times Z \times 100}{1000 \times 1.94}, \quad Z = 28.9(kg)$$

Transport cost = £20×28.9/1000 = £0.58

Treatment cost = £100×28.9/1000 = £2.89

Revenue = £28.9×0.671×0.85×400/1000 = £6.59

Therefore, NSR for zinc concentrate = £3.12/t of ore

Total NSR for perfect milling = £(5.28+3.12) = £8.40/t

2. Actual milling

Similar calculations give:

Net copper smelter return = £4.46/t ore

Net zinc smelter return = £1.71/t ore

Total net smelter return = £6.17/t ore

Therefore, overall economic efficiency = 100×6.17/8.40 = 73.5%

In recent years, attempts have been made to optimize the performance of some concentrators by controlling plant conditions to achieve maximum economic efficiency. A dilemma often facing the metallurgist on a complex flotation circuit producing more than one concentrate is: how much contamination of one concentrate by the mineral that should report to the other concentrate can be tolerated? For instance, on a plant producing separate copper and zinc concentrates, copper is always present in the zinc concentrate, as is zinc in the copper concentrate. Metals misplaced into the wrong concentrate are rarely paid for by the specialist smelter and are sometimes penalized. There is, therefore, an optimum "degree of contamination" that can be tolerated. The most important reagent controlling this factor is often the depressant, which, in this example, inhibits flotation of the zinc minerals. Increase in the addition of this reagent not only produces a cleaner copper concentrate but also tends to reduce copper recovery into this concentrate, as it also has some depressing effect on the copper minerals. The depressed copper minerals are likely to report to the zinc concentrate, so the addition rate of depressant needs to be carefully monitored and controlled to produce an optimum compromise. This should occur when the economic efficiency is maximized (Example 1.6).

1.11 SUSTAINABILITY

While making a profit is essential to sustaining any operation, sustainability has taken on more aspects in recent years, something that increasingly impacts the mining industry.

The concept of Sustainable Development arose from the environmental movement of the 1970s and was created as a compromise between environmentalists who argued that further growth and development was untenable due to depletion of resources and destructive pollution, and proponents of growth who argued that growth is required to prevent developed countries from stagnating and to allow underdeveloped countries to improve. Thus, the principles of sustainable development, that is, growth within the capacity that the earth could absorb and remain substantially unchanged, were embraced by industry. The 1987 Brundtland report, commissioned by the UN, defined sustainable development as "development that meets the needs of the present without compromising the ability of future generations to meet their own needs" (Brundtland, 1987; Anon., 2000).

Sustainable development concepts have had major impact on mining operations and projects; for example, the many safety and worker health regulations continuously promulgated since the 1960s, as well as environmental regulations controlling and limiting emissions to air, land, and sea. In fact, meeting these regulations has been a main driver of technology and operational improvement projects (van Berkel and Narayanaswamy, 2004; Marcuson, 2007).

Achieving sustainable development requires balancing economic, environmental, and social factors (Fig. 1.21). Sustainability attempts to account for the entire value of ecosystems and divides this value into three segments: direct, the value that can be generated by the animals, plants, and other resources; indirect, the value generated by items such as erosion control, water purification, and pollination; and intangible, the value to humans derived from beauty and religious/spiritual significance. Before the genesis of sustainability, economic evaluations concentrated on direct benefits undervaluing or ignoring indirect benefits. However, these indirect benefits may have significant economic value that is difficult to quantify and is only realized after the fact.

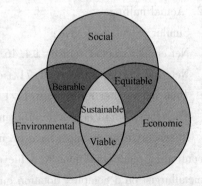

Fig. 1.21 Sustainable development: balancing economic, environmental, and social demands

The mining enterprise is directly connected to and dependent upon the earth. Mining has been a crucible for sustainability activities, especially when dealing with greenfield sites inhabited by indigenous people. Algie (2002) and Twigge-Molecey (2004) highlighted six fundamental features of mining that account for this:

(1) Metal and mineral resources are nonrenewable.

(2) Economic mineralization often occurs in remote underdeveloped areas that are high conservation zones rich in biodiversity with many sites of cultural importance to the indigenous inhabitants. The mining company becomes responsible for developing essential infrastructure. Mining activities irreversibly alter both the natural and social environments.

(3) The mining sector is diverse in size, scope, and responsibility. It comprises government and private organizations, super-major corporations and junior miners, and exploration companies.

(4) The mining industry has a poor legacy.

(5) The risks and hazards in production, use, and disposal are not well understood by the public and are often poorly communicated.

(6) In contrast to manufacturing, which primarily involves physical change, processing of metals involves chemical change, which inherently is more polluting.

In a mining endeavor, a large proportion of the benefits accrue globally while major environmental effects are felt locally. Indigenous people are closely tied to the local environment but remote from the global market exacerbating negative effects. Moreover, modern mining is not labor intensive and projects in remote locations generate jobs for only a small proportion of the aboriginal population. This may lead to an unhealthy division of wealth and status within the aboriginal community (Martin, 2005).

Sustainable development principles have dramatically impacted new projects. Extensive monitoring of baseline environmental conditions and pre-engineering studies to investigate likely impact on both the natural and human environments are mandated (Fig. 1.22). While these activities certainly identify and reduce errors made during project implementation, they exponentially in-

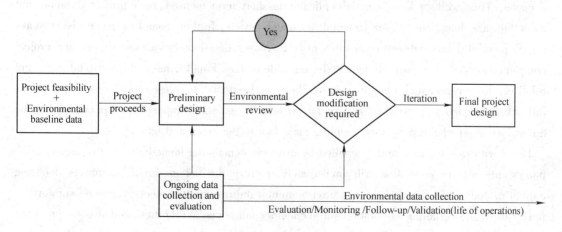

Fig. 1.22 Simplified schematic of one aspect of the iterative environmental assessment process

crease the amount of time required for completing projects. These delays may stretch project duration across several price cycles. During this time, low prices erode confidence in the worth of the project; project "champions" leave organizations, either voluntarily or involuntarily; projects are stopped and reevaluated. All of these result in more delays and additional cost.

Sustainable development has irreversibly altered the relationship between First Nations and industry (Marcuson et al., 2009). Fifty years ago, aboriginal peoples were typically ignored when creating a new project. Today, after years of advocacy, protests, and demonstrations, they have gained a place at the bargaining table, have input into the environmental permitting process, and frequently possess veto power over a new project. As a result, Impact Benefit Agreements (IBAs), designed to compensate the local population for damage to their environment and to create training, educational, and employment opportunities, are negotiated. These have succeeded in allocating a greater portion of the global benefits to the local population and in transferring social costs to the bottom line of the project. But this process has its challenges—multiparty negotiations between a company, government(s), and aboriginals take time and add costs. Moreover, governments in developing countries and aboriginal communities have different levels of sophistication in business and corporate matters, and they may have unrealistic expectations, and may be subject to exploitation by internal and external individuals and groups. Additionally, the terms of IBAs are frequently held in close confidence and thereby immune to external examination and regulation.

A modern mining project of reasonable size requires a minimum capital outlay of $1b and has an operating life of 25+ years. Costs of $4-6b are common, and time horizons of 10+years from project initiation to successful start-up are not unusual. Achieving sustainable financing to cover these costs and extended times has necessitated globalization and consolidation of the industry.

However, even mega-corporations have not successfully modulated the historic cyclic nature of the mining business. This cyclic nature presents existential challenges to those charged with sustaining mining corporations. Net Present Value, greatly influenced by price assumptions and heavily weighted to income during the first 10 years, is a major financial metric for investment decisions. Thus, while a key financial indicator is short term focused, both project duration and asset life are long, and a fundamental mismatch exists, further complicating decision making. Sequential delays between commodity price increases, decisions to increase supply, and project completion create a sluggish, fundamentally unstable system. Finally, due to high capital costs and relatively low incremental operating costs there is hesitancy to close operations when prices fall. When coupled with political and social pressure to maintain employment, operations may be maintained even when taking losses adding more fuel to the boom and bust cycle.

Faced with this situation and compelled by investors demanding immediate returns, mining companies cut costs in areas that will immediately improve the bottom line. This means deferring capital projects, negotiating delays in environmental improvement projects, reducing support for universities, and curtailing exploration and technology initiatives. While these endeavors support the long-term economic sustainability of the enterprise, curtailment and concomitant loss of expertise further exacerbate the cyclical nature of the industry.

Through regulatory and economic forces, sustainable development has become a revolutionary driver of the mining industry. These forces are increasing (Starke, 2002). Developing countries are demanding a greater portion of the economic benefits. Social and environmental sustainability become inextricably linked in the realm of water utilization and waste disposal. These present visible, immediate threats to local water supplies, the health of the natural environment, the local economies, and ways of life. Water and community health will remain flash points. To date, effective mechanisms to limit greenhouse gas emissions have not been developed. As a result, expenditures to minimize these play a minor role in new projects. But climate change is the major environmental challenge of the twenty-first century; economic and regulatory restrictions will certainly emerge as major factors in the Sustainable Development portfolio.

REFERENCES

Algie, S. H., 2002. Global materials flows in minerals processing. Green Processing 2002. AusIMM, Carlton, Australia, 39-48.

Anon., 1979. Nchanga consolidated copper mines. Eng. Min. J. 180 (Nov.), 150.

Anon., 1985a. Tin-paying the price. Min. J. (Dec), 477-479.

Anon., 1985b. Room for improvement at Chuquicamata. Min. J. (Sep), 249.

Anon., 2000. A Guide to World Resources 20002001: People and Ecosystems: The Fraying Web of Life. United Nations Development Programme, United Nations Environment Programme, World Bank and World Resources Institute, World Resources Institute, Washington, DC, USA, 1-12.

Barbery, G., 1986. Complex sulphide ores-processing options. In: Wills, B. A., Barley, R. W. (Eds.), Mineral Processing at a Crossroads: Problems and Prospects, 117. Springer, Netherlands, 157-194.

Barbery, G., 1991. Mineral Liberation, Measurement, Simulation and Practical Use in Mineral Processing. Les Editions GB, Quebec, Canada.

Baum, W., et al., 2004. Process mineralogy—a new generation for ore characterisation and plant optimisation. 2003 SME Annual Meeting and Exhibit, Denver, CO, USA, Preprint 04-012: 1-5.

Botero, A. E. C., et al., 2008. Surface chemistry fundamentals of biosorption of Rhodococcus opacus and its effect in calcite and magnesite flotation. Miner. Eng. 21 (1), 83-92.

Brierley, J. A., Brierley, C. L., 2001. Present and future commercial applications of biohydrometallurgy. Hydrometallurgy. 59 (2-3), 233-239.

Brundtland, G. (Ed.), 1987. Our Common Future: World Commission on Environment and Development. Oxford University Press, Oxford, England.

Clark, K., 1997. The business of fine coal tailings recovery. The Australian Coal Review. CSIRO Division of Coal and Energy Technology, Sydney, Australia, 3: 24-27.

Damjanovic, B., Goode, J. R. (Eds.), 2000. Canadian Milling Practice, Special vol. 49. CIM, Montre'al, QC, Canada.

Evans, C. L., et al., 2011. Application of process mineralogy as a tool in sustainable processing. Miner. Eng. 24 (12), 1242-1248.

Gaudin, A. M., 1939. Principles of Mineral Dressing. McGraw-Hill Book Company, Inc., London, England.

Gay, S. L., 2004a. A liberation model for comminution based on probability theory. Miner. Eng. 17 (4), 525-534.

Gay, S. L., 2004b. Simple texture-based liberation modelling of ores. Miner. Eng. 17 (11-12), 1209-1216.

Gilchrist, J. D., 1989. Extraction Metallurgy. Third ed. Pergamon Press, Oxford, UK.

Gottlieb, P., et al., 2000. Using quantitative electron microscopy for process mineral applications. JOM. 52 (4), 24-25.

Gray, P. M. J., 1984. Metallurgy of the complex sulphide ores. Mining Mag. Oct, 315-321.

Gu, Y., 2003. Automated scanning electron microscope based mineral liberation analysis—an introduction to JKMRC/FEI Mineral Liberation Analyser. J. Miner. Mat. Charact. Eng. 2 (1), 33-41.

Guerney, P. J., et al., 2003. Gravity recoverable gold and the Mineral Liberation Analyser. Proc. 35th Annual Meeting of the Canadian Mineral Processors Conf. CIM, Ottawa, ON, Canada, 401-416.

Hansford, G. S., Vargas, T., 2001. Chemical and electrochemical basis of bioleaching processes. Hydrometallurgy. 59 (2-3), 135-145.

Hefner, R. A., 2014. The United States of gas. Foreign Affairs. 93 (3), 9-14.

Hoal, K. O., et al., 2009. Research in quantitative mineralogy: examples from diverse applications. Miner. Eng. 22 (4), 402-408.

Iwasaki, I., Prasad, M. S., 1989. Processing techniques for difficult-totreat ores by combining chemical metallurgy and mineral processing. Miner. Process. Extr. Metall. Rev. 4 (3-4), 241-276.

Jowett, A., 1975. Formulae for the technical efficiency of mineral separations. Int. J. Miner. Process. 2 (4), 287-301.

Jowett, A., Sutherland, D. N., 1985. Some theoretical aspects of optimizing complex mineral separation systems. Int. J. Miner. Process. 14 (2), 85-109.

King, R. P., 2012. In: Schneider, C. L, King, E. A. (Eds.), Modeling and Simulation of Mineral Processing Systems, second ed. SME, Englewood, CO, USA.

Kingman, S. W., et al., 2004. Recent developments in microwaveassistedcomminution. Int. J. Miner. Process. 74 (1-4), 71-83.

Lin, C. L., et al., 2013. Characterization of rare-earth resources atMountain Pass, CA using high-resolution X-ray microtomography (HRXMT). Miner. Metall. Process. 30 (1), 10-17.

Lotter, N. O., 2011. Modern process mineralogy: an integrated multidisciplined approach to flowsheeting. Miner. Eng. 24 (12), 1229-1237.

Lynch, A. J., et al., 2007. History of flotation technology. In: Fuerstenau, M. C., et al., (Eds.), Froth Flotation: A Century of Innovation. SME, Littleton, CO, USA, 65-94.

Marcuson, S. W., 2007. SO_2 Abatement from Copper Smelting Operations: A 40 Year Perspective, vol. 3 (Book 1), The Carlos Diaz Symposium on Pyrometallurgy, Copper 2007, 39-62.

Marcuson, S. W., et al., 2009. Sustainability in nickel projects: 50 years of experience at Vale Inco, Pyrometallurgy of Nickel and Cobalt 2009. Proceedings of the 48th Annual Conference of Metallurgists of CIM, Sudbury, ON, Canada, 641-658.

Martin, D. F., 2005. Enhancing and measuring social sustainability by the minerals industry: a case study of Australian aboriginal people. Sustainable Development Indicators in the Mineral Industry, Aachen International Mining Symp. VerlagGluckauf, Essen, Germany, 663-679.

Meadows, D. H., et al., 1972. The Limits to Growth: A Report for the Club of Rome's Project on the Predicament of Mankind. Universal Books, New York, NY, USA.

Mills, C., 1978. Process design, scale-up and plant design for gravity concentration. In: Mular, A. L., Bhappu, R. B. (Eds.), Mineral Processing Plant Design. AIMME, New York, NY, USA, 404426. (Chapter 18).

Moskvitch, K., 2014. Field of dreams: the plants that love heavy metal. New Scientist. 221 (2961), 46-49.

Motl, D., et al., 2012. Advanced scanning mode for automated precious metal search in SEM (abstract only).

Annual Meeting of the Nordic Microscopy Society SCANDEM 2012, University of Bergen, Norway, 24.

Mouat, J. , 2011. Canada's mining industry and recent social, economic and business trends. In: Kapusta, J. , et al. , (Eds.), The Canadian Metallurgical & Materials Landscape 1960 to 2011. Met Soc/CIM, Westmount, Que'bec, Canada, 19-26.

Mukherjee, A. , et al. , 2004. Recent developments in processing ocean manganese nodules—a critical review. Miner. Process. Extr. Metall. Rev. 25 (2), 91-127.

Ottley, D. J. , 1991. Small capacity processing plants. Mining Mag. 165 (Nov.), 316-323.

Pease, J. D. , 2005. Fine grinding as enabling technology—The IsaMill. Proceedings of the Sixth Annual Crushing and Grinding Conference, Perth, WA, Australia, 121.

Radziszewski, P. , 2013. Energy recovery potential in comminution processes. Miner. Eng. 46-47, 83-88.

Rao, S. R. , 2006. Resource Recovery and Recycling from Metallurgical Wastes. Elsevier, Amsterdam.

Ravilious, K. , 2013. Digging downtown: the hunt for urban gold. New Scientist. 218 (2919), 40-43.

Reuter, M. A. , et al. , 2005. In: Wills, B. A. (Ed.) , the Metrics of Material and Metal Ecology: Harmonizing the Resource, Technology and Environmental Cycles. Developments in Mineral Processing. Elsevier, Amsterdam.

Ridley, J. , 2013. Ore Deposit Geology. Firsted. Cambridge University Press, Cambridge, UK.

Sassos, M. P. , 1983. Bougainville copper: tropical copper with a golden bloom feeds a 130,000-mt/d concentrator. Eng. Min. J. Oct, 56-61.

Schulz, N. F. , 1970. Separation efficiency. Trans. SME/AIME. 247 (Mar.), 81-87.

Scott, S. D. , 2001. Deep ocean mining. Geosci. Canada. 28 (2), 87-96.

Smith, R. W. , et al. , 1991. Mineral bioprocessing and the future. Miner. Eng. 4 (7-11), 1127-1141.

Smythe, D. M. , et al. , 2013. Rare earth element deportment studies utilising QEMSCAN technology. Miner. Eng. 52, 52-61.

Starke, J. (Ed.), 2002. Breaking New Ground: Mining, Minerals and Sustainable Development: The Report of the MMSD Project. Mining, Minerals, and Sustainable Development Project, EarthscanPublications Ltd. , London, England.

Tan, L. , Chi-Lung, Y. , 1970. Abundance of the chemical elements in The Earth's crust. Int. Geol. Rev. 12 (7), 778-786.

Taylor, S. R. , 1964. Abundance of chemical elements in the continental crust: a new table. Geochim. Cosmochim. Acta. 28 (8), 1273-1285.

Thomas, K. G. , et al. , 2014. Project execution & cost escalation in the mining industry. Proc. 46th Annual Meeting of the Canadian Mineral Processors Conf. CIM, Ottawa, Canada, 105-114.

Twigge-Molecey, C. , 2004. Approaches to plant design for sustainability. Green Processing 2004. AusIMM, Fremantle, WA, Australia, 47-52.

USGS, 2014. , http://minerals. usgs. gov/. (Accessed Aug. 2014.).

VanBerkel, R. , Narayanaswamy, V. , 2004. Sustainability as a framework for innovation in minerals processing. Green Processing 2004. AusIMM, Fremantle, WA, Australia, 197-205.

Wills, B. A. , Atkinson, K. , 1991. The development of minerals engineering in the 20th century. Miner. Eng. 4 (7-11), 643-652.

Wills, B. A. , Atkinson, K. , 1993. Some observations on the fracture and liberation of mineral assemblies. Miner. Eng. 6 (7), 697-706.

Chapter 2 Ore Handling

2.1 INTRODUCTION

Ore handling is a key function in mining and mineral processing, which may account for 30%-60% of the total delivered price of raw materials. It covers the processes of transportation, storage, feeding, and washing of the ore en route to, or during, the various stages of treatment in the mill.

Since the physical state of ores in situ may range from friable, or even sandy material, to monolithic deposits with the hardness of granite, the methods of mining and provisions for the handling of freshly excavated material will vary widely. Ore that has been well fragmented can be transported by trucks, belts, or even by sluicing, but large lumps of hard ore may need secondary blasting.

Developments in nonelectric millisecond delay detonators and plastic explosives have resulted in more controllable primary breakage and easier fragmentation of occasional overly-large lumps. At the same time, crushers have become larger and lumps up to 2m in size can now be fed into some primary units.

Ores are by and large heterogeneous in nature. The largest lumps blasted from an open pit operation may be over 1.5m in size. The fragmented ore from a blast is loaded directly into trucks, holding up to 400t of ore in some cases, and is transported directly to the primary crushers. Storage of such ore is not always practicable, due to its wide particle size range which causes segregation during storage, the fines working their way down through the voids between the larger particles. Extremely coarse ore is sometimes difficult to start moving once it has been stopped. Sophisticated storage and feed mechanisms are therefore often dispensed with, the trucks depositing their loads directly on the grizzly feeding the primary crusher.

The operating cycle of an underground mine is complex. Drilling and blasting are often performed in one or two shifts; the blasted ore is hoisted to the surface during the next couple of shifts. The ore is transported through the passes via chutes and tramways and is loaded into skips, holding as much as 30t of blasted ore, to be hoisted to the surface. Large boulders are often broken up underground by primary rock breakers in order to facilitate loading and handling at this stage. The ore, on arrival at the surface, having undergone some initial crushing, is easier to handle than that from an open pit mine. The storage and feeding is usually easier, and indeed essential, due to the intermittent arrival of skips at the surface.

2.2 THE REMOVAL OF HARMFUL

MATERIALS: Ore entering the mill from the mine (run-of-mine ore) normally contains a small proportion of material which is potentially harmful to the mill equipment and processes.

For instance, large pieces of iron and steel broken off from mine machinery can jam in the crushers. Wood is a major problem in many mills as it is ground into a fine pulp and causes choking or blocking of screens, flotation cell ports, etc. Wood pulp may also consume flotation reagents by absorption, which reduces mineral floatability.

Clays and slimes adhering to the ore are also harmful as they hinder screening, filtration, and thickening, and again may consume flotation reagents.

All these tramp materials must be removed as far as possible at an early stage in treatment. Removal by hand (hand sorting) from conveyor belts has declined with the development of mechanized methods of dealing with large tonnages, but it is still used when plentiful cheap labor is available.

Skips, ore bins, and mill equipment can be protected from large pieces of "tramp" iron and steel, such as rockbolts and wire meshes, by electromagnets suspended over conveyor belts (guard magnets) (Fig. 2.1). The magnets are generally installed downstream of the primary crusher to protect skips and ore bins. These powerful electromagnets can pick up large pieces of iron and steel travelling over the belt. They may operate continuously (as shown) or be stationary and, at intervals, are swung away from the belt and unloaded when the magnetic field is removed. Guard magnets, however, cannot be used to remove tramp iron from magnetic ores, such as those containing magnetite, nor will they remove nonferrous metals or nonmagnetic steels from the belt. Metal detectors, which measure the electrical conductivity of the material being conveyed, can be fitted over or around conveyor belts. The electrical conductivity of ores is much lower than that of metals, and fluctuations in electrical conductivity in the conveyed material can be detected by measuring the change that tramp metal causes in a given electromagnetic field. When a metal object triggers an alarm, the belt automatically stops and the object can be removed. With nonmagnetic ores it is advantageous to precede the metal detector with a guard magnet, which will remove the ferromagnetic tramp metals and thus minimize belt stoppages.

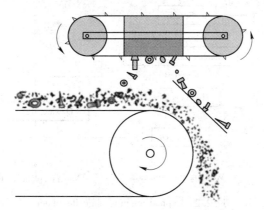

Fig. 2.1 Conveyor guard magnet (see also Chapter 6)

Large pieces of wood that have been flattened by passage through a primary crusher can be removed by passing the ore feed over a vibrating scalping screen. Here, the apertures of the screen are slightly larger than the maximum size of particle in the crusher discharge, allowing the ore to fall through the apertures and the flattened wood particles to ride over the screen and be collected separately. (On cold nights the collected wood might find a use.)

Wood can be further removed from the pulp discharge from the grinding mills by passing the pulp through a fine screen. Again, while the ore particles pass through the apertures, the wood collects on top of the screen and can be periodically removed.

Washing of run-of-mine ore can be carried out to facilitate sorting by removing obscuring dirt from the surfaces of the ore particles. However, washing to remove very fine material, or slimes, of little or no value is more important.

Washing is normally performed after primary crushing as the ore is then of a suitable size to be passed over washing screens. It should always precede secondary crushing as slimes severely interfere with this stage. The ore is passed through high-pressure jets of water on mechanically vibrated screens. The screen undersize product is usually directed to the grinding mills and thus the screen apertures are usually of similar size to the particles in the feed to the grinding mills.

Ore washing is sometimes assisted by adding scrubbers in the circuit. Scrubbers are designed to clean crushed ore, sand, and gravel, but they can also upgrade an ore by removing soft rock by attrition. Scrubbers are self-aligning, steel trunnions supported on flanged railroad type bearings, and driven by a saddle drive chain.

In the circuit shown in Fig. 2.2, material passing over the screen, that is, washed ore, is transported to the secondary crushers. Material passing through the screens is classified into coarse and fine fractions by a mechanical classifier or hydrocyclone (Chapter 3), or both. It may be beneficial to classify initially in a mechanical classifier as this is more able to smooth out fluctuations in flow than is the hydrocyclone and it is better suited to handling coarse material.

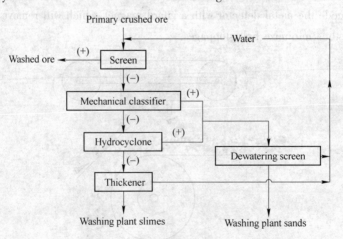

Fig. 2.2 Typical washing plant flowsheet

The coarse product from the classifier, designated "washing plant sands", is either routed direct to the grinding mills or is dewatered over vibrating screens before being sent to mill storage. A considerable load, therefore, is taken off the dry crushing section.

The fine product from classification, that is, the "slimes", may be partially dewatered in shallow large diameter settling tanks known as thickeners, and the thickened pulp is either pumped to tailings disposal or, if containing values, pumped direct to the concentration process, thus removing load from the grinding section. In Fig. 2.2, the thickener overflows are used to feed the high-pressure washing sprays on the screens. Water conservation in this manner is practiced in most mills.

Wood pulp may again be a problem in the above circuit, as it will tend to float in the thickener, and will choke the water spray nozzles unless it is removed by retention on a fine screen.

2.3 ORE TRANSPORTATION

In a mineral processing plant, operating at the rate of 400000t/d, this is equivalent to about 28t of solid per minute, requiring up to 75m³/min of water. It is there-fore important to operate with the minimum upward or horizontal movement and with the maximum practicable pulp density in all of those stages subsequent to the addition of water to the system. The basic philosophy requires maximum use of gravity and continuous movement over the shortest possible distances between processing units.

Dry ore can be moved through chutes, provided they are of sufficient slope to allow easy sliding and sharp turns are avoided. Clean solids slide easily on a 15°~25° steel-faced slope, but for most ores, a 45°~55° working slope is used. The ore may be difficult to control if the slope is too steep.

The belt conveyor system is the most effective and widely used method of handling loose bulk materials in mining and mineral processing industries. In a belt conveyor system, the belt is a flexible and flat loop, mounted over two pulleys, one of which is connected to a drive to provide motion in one direction. The belt is tensioned sufficiently to ensure good grip with the drive pulley, and is supported by a structural frame, with idlers or slider bed in between the pulleys. Belts today have capacities up to 40000t/h (Alspaugh, 2008) and single flight lengths exceeding 15000m, with feasible speeds of up to 10m/s.

The standard rubber conveyor belt has a foundation, termed a carcass, of sufficient strength to withstand the belt tension, impact, and strains due to loading. This foundation can be single-ply or multi-ply (Fig. 2.3) and is made of cotton, nylon, leather, plastic, steel fabric, or steel cord. The foundation is bound together with a rubber matrix and completely covered with a layer of vulcanized rubber. The type of vulcanized rubber cover may vary depending upon the ore properties and operational conditions (e.g., abrasiveness of ore, powder or lump material, temperature) (Ray, 2008).

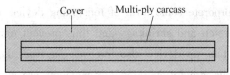

Fig. 2.3 Construction of a multi-ply belt (in cross section)

The carrying capacity of the belt is increased by passing it over troughing idlers. These are support rollers set normal to the travel of the belt and inclined upward from the center so as to raise the edges and give it a trough-like profile. There may be three or five in a set and they will be rubber-coated under a loading point, so as to reduce the wear and damage from impact. Spacing along the belt is at the maximum interval that avoids excessive sag. The return belt is supported by horizontal straight idlers that overlap the belt by a few inches at each side.

The idler dimensions and troughing angle are laid down in BIS in IS 8598:1987(2). The diameters of carrying and return idlers range from ca. 63 to 219mm. Idler length may vary from 100mm to 2200mm. Controlling factors for idler selection may include unit weight, lump size, and belt speed. The length of an idler is proportional to its diameter. The troughing idler sets are installed with a troughing angle ranging $15° \sim 50°$. Idler spacing on the loaded side is a function of unit weight of material and belt width (Ray, 2008).

To control belt wandering (lateral movement), either training idlers are installed normal to the direction of the belt motion or wing idlers are installed at a forward angle. Sensors and electronic devices are also used to detect wandering and keep the belt on track (Anon., 2014).

The pulleys of belt conveyors are manufactured from rolled steel plates or from cast iron. The pulley drum is keyed to the steel shaft and the finished dimensions are machined. Generally, crowning is done to the pulley face to decrease belt wandering at the pulley. The length of the pulley is $100 \sim 200$mm more than the belt width (Ray, 2008). Inducing motion without slipping requires good contact between the belt and the drive pulley. This may not be possible with a single 180° turn over a pulley (wrap angle, α), and some form of "snubbed pulley" drive (Fig. 2.4 (a)) or "tandem" drive arrangement (Fig. 2.4 (b) and (c)) may be more effective. The friction can also be provided by embedded grooves or covering the pulleys with rubber, polyurethane, or ceramic layer of thickness $6 \sim 12$mm. The layer can also be engraved with patterned grooves for increased friction and better drainage when dealing with wet material (Ray, 2008).

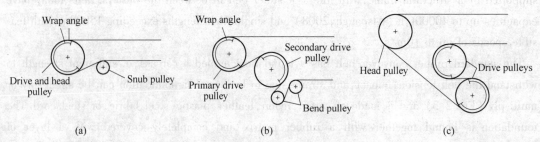

Fig. 2.4 Conveyor-belt drive arrangements
(a) single drive, (b) tandem drive for underground mines, and (c) tandem drive for open pit mines

The belt system must incorporate some form of tensioning device, also known as belt take-up load, to adjust the belt for stretch and shrinkage and thus prevent undue sag between idlers, and slip at the drive pulley. The take-up device also removes sag from the belt by developing tensile stress in the belt. In most mills, gravity-operated arrangements are used, which adjust the tension continuously (Fig. 2.5). A screw type take-up loading system can be used instead of gravity for tensioning the conveyor (Ray,2008). Hydraulics have also been used extensively, and when more refined belt-tension control is required, especially in starting and stopping long conveyors, load-cell-controlled electrical tensioning devices are used.

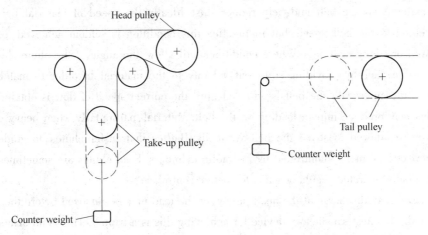

Fig. 2.5 Conveyor-belt tensioning systems

Advances in control technology have enhanced the reliability of belt systems by making possible a high degree of fail-safe automation. A series of belts should incorporate an interlock system such that failure of any particular belt will automatically stop preceding belts. Interlock with devices being fed by the belt is important for the same reasons. It should not be possible to shut down any machine in the system without arresting the feed to the machine at the same time and, similarly, motor failure should lead to the automatic tripping (stopping) of all preceding belts and machines. Similarly, the belt start up sequence of conveyor systems is fixed such that the last conveyor should start first followed by the second to last and so on. Sophisticated electrical, pneumatic, and hydraulic circuits have been widely employed to replace all but a few manual operations.

Several methods can be used to minimize loading shock on the belt. A typical arrangement is shown in Fig. 2.6 where the fines are screened onto the belt first and provide a cushion for the lar-

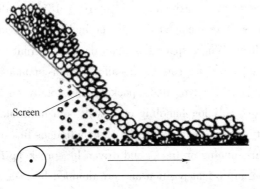

Fig. 2.6 Belt-loading system

ger pieces of rock. The impact loading on a belt can also be reduced by increasing the number of idlers at the loading points; such idlers are called impact idlers.

Feed chutes must be designed to deliver the bulk of the material to the center of the belt at a velocity close to that of the belt. Side boards or skirt plates with a length of 2~3m are installed to guide the material on the belt and help reduce dust. Ideally the speed of material being placed should be equal to the belt speed, but in practice this condition is seldom achieved, particularly with wet sand or sticky materials. Where conditions will allow, the angle of the chute should be as great as possible, thereby providing sufficient velocity to the material in order to match the belt speed. The chute angle with the belt is adjusted until the correct speed of flow is obtained. Higher chute angles may produce impact loading on the belt. Material, particularly when heavy or lumpy, should never be allowed to strike the belt vertically. Baffles in transfer chutes, to guide material flow, are now often remotely controlled by hydraulic cylinders. Feed chutes are sometimes installed with hydraulically operated regulator gates for better control.

The conveyor may discharge at the head pulley, or the load may be removed before the head pulley is reached. The most satisfactory device for achieving this is a tripper. This is an arrangement of pulleys in a frame by which the belt is raised and doubled back so as to give it a localized discharge point. The frame is usually mounted on wheels, running on tracks, so that the load can be delivered at several points, over a long bin or into several bins. The discharge chute on the tripper can deliver to one or both sides of the belt. The tripper may be moved manually, by head and tail ropes from a reversible hoisting drum, or by a motor. It may be automatic, moving backward and forward under power from the belt drive. A plough can also be used to discharge the material. A plough is a v-shaped, rubber tipped blade extending along the width of the conveyor at an angle of 60°. A troughed conveyor is made flat by passing over a slider, at such a discharge point. Cameras allow the level of bin filling to be monitored.

Shuttle belts are reversible self-contained conveyor units mounted on carriages, which permit them to be moved lengthwise to discharge to either side of the feed point. The range of distribution is approximately twice the length of the conveyor. They are often preferred to trippers for permanent storage systems because they require less head room and, being without reverse bends, are much easier on the belt.

Belt conveyors can operate at a maximum angle of 10°~18° depending on the material being conveyed. Beyond the range of recommended angle of incline, the material may slide down the belt, or it may topple on itself. Where space limitation does not permit the installation of a regular belt conveyor, steep angle conveying can be installed which includes gravity bucket elevators (Fig. 2.7), molded cleat belts, fin type belts, pocket belts, totally enclosed belts, and sandwich belts. All these methods give only low handling rates with both horizontal conveying and elevating of the material. The gravity bucket elevators consist of a continuous line of buckets attached by pins to two endless roller chains running on tracks and driven by sprockets. The buckets are pivoted so that they always remain in an upright position and are dumped by means of a ramp placed to engage a shoe on the bucket, thus turning it into the dumping position.

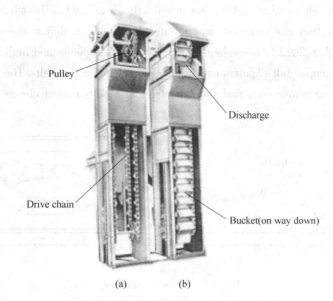

Fig. 2.7 Gravity bucket elevator (a) rear view without buckets, and (b) return (empty) buckets

Pipe conveyors were developed in Japan and are now being marketed throughout the world. This belt conveyor, after being loaded, is transformed into a pipe by arranging idlers. Five or six idlers are used to achieve the belt wrapping. The conveyor is unwrapped at the discharge point using idler arrangement and the belt is then passed over the head pulley in the conventional manner (Fig. 2.8). The pipe conveyors are clean and environment friendly, especially when it comes to transporting hazardous material. This conveyor is compact and does not require a hood. It can efficiently negotiate horizontal curves with a short turning radius (McGuire, 2009).

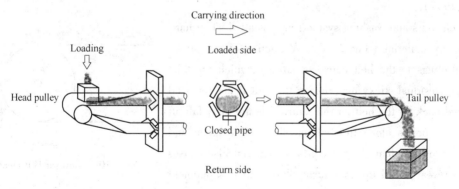

Fig. 2.8 Pipe conveyor; note the return side can also be loaded

Sandwich conveyor systems can be used to transport solids at steep inclines from 30° to 90°. The material being transported is "sandwiched" between two belts that hold the material in position and prevent it from sliding, rolling back or leaking back down the conveyor even after the conveyor has stopped or tripped. As pressure is applied to material to hold it in place, it is important that the material has a reasonable internal friction angle. Pressure is applied through roller arrangements, special lead-filled belts, belt weights, and tension in the belt (Fig. 2.9). The advantage of sandwich belt conveyors is that they can transport material at steep angles at similar speeds to conventional belt conveyors (Walker, 2012). These belts can achieve high capacity and high lift. They can also accommodate large lumps. Belt cleaners can be used for cleaning the belts. The disadvantages include extra mechanical components and maintenance. These belts cannot elevate fine materials effectively (Anon., 2014).

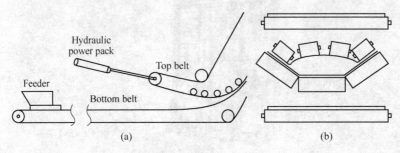

Fig. 2.9 Sandwich conveyor
(a) side view, and (b) sectional view

Screw conveyors, also termed auger conveyors, are another means of transporting dry or damp particles. The material is pushed along a trough by the rotation of a helix, which is mounted on a central shaft. The action of the screw conveyor allows for virtually any degree of mixing of different materials and allows for the transportation of material on any incline from the horizontal to vertical. The screw conveyors are also capable of moving nonfree flowing and semi-solid materials (slurries) as well (Bolat and Bogoclu, 2012; Patel et al., 2013). The main limitation of screw conveyors is the feed particle size and capacity, which has a maximum of ca. 300m^3/h (Perry and Green, 1997).

Belt cleaners and washing systems are installed when handling sticky material (Fig. 2.10). A fraction of the sticky material clings to the belt conveyor surface, which must be removed. Residual sticky material on the belt if not removed is carried back by the belt on the return side and may fall off at different points along the belt causing maintenance and housekeeping problems. The carry-back material also causes excessive wear, build up on return idlers, and misalignment

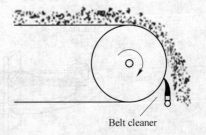

Fig. 2.10 Conveyor belt cleaner

and damage to the belt due to the accumulation of material. Belt cleaners and washing systems are generally installed near the discharge point (Anon., 2014).

Hydraulic transport of the ore stream normally takes over from dry transportation when ore is mixed with water, which is the grinding stage in most mills. Pulp may be made to flow through open launders by gravity in some cases. Launders, also termed flumes, are gently sloping troughs of rectangular, triangular, or semicircular section, in which the solid is carried in suspension, or by sliding or rolling. In mills, only fine sized ores (−5mm) are transported. The slope must increase with particle size, with the solid content of the suspension, and with specific gravity of the solid. The effect of depth of water is complex: if the particles are carried in suspension, a deep launder is advantageous because the rate of solid transport is increased. If the particles are carried by rolling, a deep flow may be disadvantageous.

In plants of any size, the pulp is moved through piping via centrifugal pumps. Pipelines should be as straight as possible to prevent abrasion at bends. Abrasion can be reduced by using liners inside the pipelines. The use of oversize pipe is to be avoided whenever slow motion might allow the solids to settle and hence choke the pipe. The factors involved in pipeline design and installation are complex and include the solid-liquid ratio, the average pulp density, the density of the solid constituents, the size analysis and particle shape, and the fluid viscosity (Loretto and Laker, 1980).

Centrifugal pumps are cheap in capital cost and maintenance and occupy little space (Wilson, 1981; Pearse, 1985). Single-stage pumps are normally used, lifting up to 30m and in extreme cases 100m. The merits of centrifugal pumps include no drive seals, very little friction is produced in the pump, and almost no heat is transferred from the motor. Also, the centrifugal pump is not prone to breakage due to the coupling arrangement of the pump and motor (Wilson et al., 2006; Gulich, 2010). Their main disadvantage is the high velocity produced within the impeller chamber, which may result in serious wear of the impeller and chamber itself, especially when coarse sand is being pumped.

2.4 ORE STORAGE

The necessity for storage arises from the fact that different parts of the operation of mining and milling are performed at different rates, some being intermittent and others continuous, some being subject to frequent interruption for repair and others being essentially batch processes. Thus, unless reservoirs for material are provided between the different steps, the whole operation is rendered spasmodic and, consequently, uneconomical. Ore storage is a continuous operation that runs 24h a day and 7 days a week. The type and location of the material storage depends primarily on the feeding system. The ore storage facility is also used for blending different ore grades from various sources.

The amount of storage necessary depends on the equipment of the plant as a whole, its method of operation, and the frequency and duration of regular and unexpected shutdowns of individual units.

For various reasons, at most mines, ore is hoisted for only a part of each day. On the other hand, grinding and concentration circuits are most efficient when running continuously. Mine operations are more subject to unexpected interruption than mill operations, and coarse-crushing machines are more subject to clogging and breakage than fine crushers, grinding mills, and concentration equipment. Consequently, both the mine and the coarse-ore plant should have a greater hourly capacity than the fine crushing and grinding plants, and storage reservoirs should be provided between them. Ordinary mine shutdowns, expected or unexpected, will not generally exceed a 24h duration, and ordinary coarse-crushing plant repairs can be made within an equal period if a good supply of spare parts is kept on hand. Therefore, if a 24h supply of ore that has passed the coarse-crushing plant is kept in reserve ahead of the mill proper, the mill can be kept running independent of shutdowns of less than a 24h duration in the mine and coarse-crushing plant. It is wise to provide for a similar mill shutdown and, in order to do this, the reservoir between coarse-crushing plant and mill must contain at all times unfilled space capable of holding a day's tonnage from the mine. This is not economically possible, however, with many of the modern very large mills; there is a trend now to design such mills with smaller storage reservoirs, often supplying less than a two-shift supply of ore, the philosophy being that storage does not do anything to the ore, and can, in some cases, has an adverse effect by allowing the ore to oxidize. Unstable sulfides must be treated with minimum delay, the worst case scenario being self-heating with its attendant production and environmental problems (Section 2.6). Wet ore cannot be exposed to extreme cold as it will freeze and become difficult to move.

Storage has the advantage of allowing blending of different ores so as to provide a consistent feed to the mill. Both tripper and shuttle conveyors can be used to blend the material into the storage reservoir. If the units shuttle back and forth along the pile, the materials are layered and mix when reclaimed. If the units form separate piles for each quality of ore, a blend can be achieved by combining the flow from selected feeders onto a reclaim conveyor.

Depending on the nature of the material treated, storage is accomplished in stockpiles, bins, or tanks. Stockpiles are often used to store coarse ore of low value outdoors. In designing stockpiles, it is merely necessary to know the angle of repose of the ore, the volume occupied by the broken ore, and the tonnage. The stockpile must be safe and stable with respect to thermal conductivity, geomechanics, drainage, dust, and any radiation emission. The shape of a stockpile can be conical or elongated. The conical shape provides the greatest capacity per unit area, thus reduces the plant footprint. Material blending from a stockpile can be achieved with any shape but the most effective blending can be achieved with elongated shape.

Although material can be reclaimed from stockpiles by front-end loaders or by bucket-wheel reclaimers, the most economical method is by the reclaim tunnel system, since it requires a minimum of manpower to operate (Dietiker, 1980). It is especially suited for blending by feeding from any combination of openings. Conical stock-piles can be reclaimed by a tunnel running through the center, with one or more feed openings discharging via gates, or feeders, onto the reclaim belt. Chain scraper reclaimers are the alternate device used, especially for the conical stock pile. The amount of reclaimable material, or the live storage, is about 20% ~ 25% of the total (Fig. 2.11). Elongated stockpiles are reclaimed in a similar manner, the live storage being 30% ~ 35% of the total (Fig. 2.12).

For continuous feeding of crushed ore to the grinding section, feed bins are used for transfer of the coarse mate-rial from belts and rail and road trucks. They are made of wood, concrete, or steel. They must be easy to fill and must allow a steady fall of the ore through to the discharge gates with no "hanging up" of material or opportunity for it to segregate into coarse and fine frac-

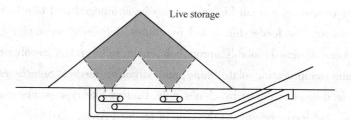

Fig. 2.11　Reclamation from conical stock pile

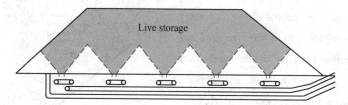

Fig. 2.12　Reclamation from elongated stock pile

tions. The discharge must be adequate and drawn from several alternative points if the bin is large. Flat-bottom bins cannot be emptied completely and retain a substantial tonnage of dead rock. This, however, provides a cushion to protect the bottom from wear, and such bins are easy to construct. This type of bin, however, should not be used with easily oxidized ore, which might age dangerously and mix with the fresh ore supply. Bins with sloping bottoms are better in such cases.

Pulp storage on a large scale is not as easy as dry ore storage. Conditioning tanks are used for storing suspensions of fine particles to provide time for chemical reactions to proceed. These tanks must be agitated continuously, not only to provide mixing but also to prevent settlement and choking up. Surge tanks are placed in the pulp flow-line when it is necessary to smooth out small operating variations of feed rate. Their content can be agitated by stirring, by blowing in air, or by circulation through a pump.

2.5　FEEDING

Feeders are necessary whenever it is desired to deliver a uniform stream of dry or moist ore, since such ore will not flow evenly from a storage reservoir of any kind through a gate, except when regulated by some type of mechanism.

Feeding is essentially a conveying operation in which the distance travelled is short and in which close regulation of the rate of passage is required. Where succeeding operations are at the same rate, it is unnecessary to inter-pose feeders. Where, however, principal operations are interrupted by a storage step, it is necessary to provide a feeder. Feeders also reduce wear and tear, abrasion, and segregation. They also help in dust control and reduce material spillage. Feeder design must consider desired flow rates, delivery of a stable flow rate, the feeding direction, and particle size range of feed to be handled (Roberts, 2001).

A typical feeder consists of a small bin, which may be an integral part of a large bin, with a gate and a suitable conveyor. The feeder bin is fed by chutes, delivering ore under gravity. Feeders of many types have been designed, notably apron, belt, chain, roller, rotary, revolving disc, drum, drag scraper, screw, vane, reciprocating plate, table, and vibrating feeders. Sometimes feeders are not used and instead feeding is achieved by chutes only. Factors like type of material to be handled, the storage method, and feed rate govern the type of feeder (Anon., 2014).

In the primary crushing stage, the ore is normally crushed as soon as possible after its arrival. Many under-ground mines have primary crushers underground to reduce ore size and improve hoisting efficiency. Skips, trucks, and other handling vehicles are intermittent in arrival, whereas the crushing section, once started, calls for steady feed. Surge bins provide a convenient holding arrangement able to receive all the intermittent loads and to feed them steadily through gates at controllable rates. The chain-feeder (Fig. 2.13) is sometimes used for smooth control of bin discharge.

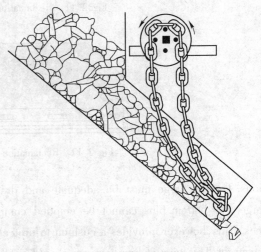

Fig. 2.13 Side view of a chain-feeder

The chain-feeder consists of a curtain of heavy loops of chain, lying on the ore at the outfall of the bin at approximately the angle of repose. The rate of feed is controlled automatically or manually by the chain sprocket drive such that when the loops of chain move, the ore on which they rest begins to slide.

Primary crushers depend for normal operation on the fact that broken rock contains a certain voidage (space between particles). If all the feed goes to a jaw crusher without a preliminary removal of fines, there can be danger of choking when there has been segregation of coarse and fine material in the bin. Such fines could pass through the upper zones of the crusher and drop into the finalizing zone and fill the voids. Should the bulk arriving at any level exceed that departing, it is as though an attempt is being made to compress solid rock. This choking, that is, packing of the crushing chamber (or "bogging"), is just as serious as tramp iron in the crusher and likewise can cause major damage. It is common practice, therefore, to "scalp" the feed to the crusher, heavy-duty screens known as grizzlies normally preceding the crushers and removing fines.

Primary crusher feeds, which scalp and feed in one operation, have been developed, such as the vibrating grizzly feeder. The elliptical bar feeder (Fig. 2.14) consists of elliptical bars of steel which form the bottom of a receiving hopper and are set with the long axes of the ellipses in alternate vertical and horizontal positions. Material is dumped directly onto the bars, which rotate in the same direction, all at the same time, so that the spacing remains constant. As one turns down, the succeeding one turns up, imparting a rocking, tumbling motion to the load. This works to loosen the fines, which sift through the load directly on to a conveyor belt, while the oversize is moved forward to deliver to the crusher. This type of feeder is probably better suited to handling high clay or wet materials such as laterite, rather than hard, abrasive ores.

The apron feeder (Fig. 2.15) is one of the most widely used feeders for handling coarse ore, especially jaw crusher feed. The overlapping metal plates or pans mounted on strands of conveyor chains convey the material (Anon.,2014). It is ruggedly constructed, consisting of a series of high carbon or manganese steel pans, bolted to strands of heavy-duty chain, which run on steel sprockets. The rate of discharge is controlled by varying the speed or the height of the ribbon of ore by means of an adjustable gate. It can handle abrasive, heavy, and lumpy materials (Anon.,2014).

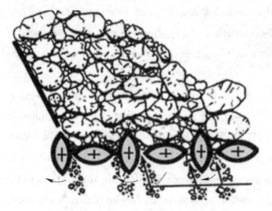

Fig. 2.14 Cross section of elliptical bar feeder

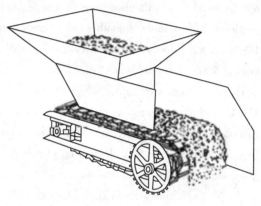

Fig. 2.15 Apron feeder

Apron feeders are often preferred to reciprocating plate feeders which push forward the ore lying at the bottom of the bin with strokes at a controllable rate and amplitude, as they require less driving power and provide a steadier, more uniform feed.

Belt feeders are essentially short belt conveyors, used to control the discharge of material from inclined chutes. The belt is flat and is supported by closely spaced idlers. They frequently replace apron feeders for fine ore and are increasingly being used to handle coarse, abrasive, friable primary crushed ore. Compared with apron feeders, they require less installation height, cost substantially less, and can be operated at higher speeds.

2.6 SELF-HEATING OF SULFIDE MINERALS

Self-heating is a problem associated with many materials that affects how they are handled, stored, and transported (Quintiere et al.,2012). Self-heating is also referred to as spontaneous heating and pyrophoric behavior and results when the rate of heat generation (due to oxidation) exceeds the rate of heat dissipation. In the minerals industry, environmental effects of self-heating for coals are well documented, from the production of toxic fumes (CO, NO_x, SO_2) and greenhouse gases (CH_4, CO_2), to the contamination of runoff water (Kim,2007;Stracher,2007). There is also growing concern over self-heating of sulfides as regulations for shipping tighten (Anon.,2011).

Many base metals occur in nature as mineral sulfides, a form which has made their extraction, concentration, and conversion into metals a challenge, but a challenge that has been successfully met by technologies such as flotation, leaching, and autogenous smelting. The propensity of sulfur-containing materials to oxidize is largely the reason for the successful extraction of these metals, as well as the source of some of the associated problems of base metal processing. These problems include acid rock (acid mine) drainage, dust explosions, and self-heating of ores, concentrates, waste rock, tailings, and mine paste fill.

Heating may occur when the sulfide material is contained or piled in sufficient quantity (i.e., reducing the heat dissipation rate), with both oxygen (air) and some moisture present (ca. 3%~8% by weight). If conditions are favorable, and this includes long storage times, presence of fine particles, high relative humidity, and temperatures exceeding 30℃, heating can proceed beyond 100℃, at which point SO_2 gas begins to evolve and may continue to drive temperatures well in excess of 400℃ (Rosenblum et al., 2001).

Fig. 2.16 shows examples of waste rock (a) and concentrate (b) that have heated beyond 100℃ leading to evolution of SO_2. "Hot muck" underground at the Sullivan leadzinc mine in British Columbia, Canada, made the cover of the CIM Bulletin magazine (June 1977). That mine first reported issues with self-heating of ore in 1926 (O'Brien and Banks, 1926), illustrating that the issues associated with sulfide self-heating have been around for considerable time. The sinking of the N. Y. K. Line's SS Bokuyo Maru in 1939 was attributed to spontaneous combustion of copper concentrate (Kirshenbaum, 1968). The consequences are rarely so dramatic but can result in significant storage and transportation issues that may threaten infrastructure and the workplace environment.

Fig. 2.16 Examples of self-heating of sulfides

(a) steam (and SO_2) emanating from sulfide ore waste dump, and (b) high temperature

and SO_2 emanating from stockpiled copper sulfide concentrate

((a) Courtesy T. Krolak; (b) courtesy F. Rosenblum)

Dealing with materials that have the potential to self-heat requires an understanding of the material reactivity and a proactive risk management approach (Rosenblum, et al., 2001). A variety of single-stage testing methods are in use for different materials with potential for self-heating (e.g., coal, wood chips, powdered milk). However, mineral sulfides require a two-stage assessment: one that mimics weathering (i.e., oxidation) at near ambient conditions and where elemental sulfur is created, followed by a higher temperature stage above 100℃ to assess the impact of the weathering stage and where the elemental sulfur is oxidized to form SO_2 (Rosenblum et al., 2014).

It is thought that the reactions governing self-heating are electrochemical as well as thermodynamic in origin (Payant et al., 2012; Somot and Finch, 2010). Pure sulfide minerals do not readily self-heat, the exception being pyrrhotite ($Fe_{1-x}S$), likely due to its nonstochiometric excess of sulfur. Payant et al. (2012) have reported that a difference in the electrochemical rest potential between minerals in a binary mixture needs to exceed 0.2V in order for self-heating to proceed. From Table 2.1, this means the pyrite-galena mix will self-heat, and that pyrite will accelerate self-heating of pyrrhotite, as observed experimentally (Payant et al., 2012).

Table 2.1 Rest Potential Values of Some Sulfide Minerals

Mineral	Formula[①]	Rest Potential (vs. SHE)/V
Pyrite	FeS_2	0.66
Chalcopyrite	$CuFeS_2$	0.56
Sphalerite	ZnS	0.46
Pentlandite	$(Fe,Ni)_9S_8$	0.35
Pyrrhotite	$Fe_{1-x}S$	0.31
Galena	PbS	0.28

Source: From Payant et al. (2012).

[①]Nominal formula, natural samples can vary.

Mitigation strategies used to control the risk of self-heating include controlling pyrrhotite content to below 10%wt, monitoring for hot-spots with infrared thermal detectors, blending any hot material with cooler material, "blanketing" with CO_2 (in ships' holds, and storage sheds), drying to below 1%wt moisture, and sealing with plastic (e.g., shipping concentrate in tote bags) to eliminate oxygen. The addition of various chemical agents to act as oxidation inhibitors is reportedly also practiced.

REFERENCES

Alspaugh, M., 2008. Bulk Material Handling by Conveyor Belt 7. SME, Littleton, CO., USA.

Anon, 2011. The International Maritime Solid Bulk Cargoes (IMSBC) Code adopted by Resolution MSC.268 (85) (Marine Safety Committee). DNV Managing Risk, UK.

Anon. (2014). Belt conveyors for bulk materials. Report by members of CEMA Engineering Conference, Conveyor Equipment Manufacturers Association (CEMA), and CBI Pub. Co., Boston, MA, USA.

Bolat, B., Bŏgoc, lu, M.E. (2012). Increasing of screw conveyor capacity. Proceedings of the 16th International Research/Expert Conference, Trends in the Development of Machinery and Associated Tech., TMT, Dubai, UAE, 515-518.

Dietiker, F.D., 1980. Belt conveyor selection and stockpiling and reclaiming applications. In: Mular, A.L., Bhappu, R.B. (Eds.), Mineral Processing Plant Design, second ed. SME, New York, NY, USA, 618-635. (Chapter 30).

Gülich, J. F. , 2010. Centrifugal Pumps. Second ed. Springer, New York, NY, USA. Kim, A. G. , 2007. Greenhouse gases generated in underground coal-mine fires. Rev. Eng. Geol. 18, 1-13.

Kirshenbaum, N. W. , 1968. Transport and Handling of Sulphide Concentrates: Problems and Possible Improvements. Second ed. Technomic Publishing Company Inc. , Pennsylvania, PA, USA.

Loretto, J. C. , Laker, E. T. , 1980. Process piping and slurry transportation. In: Mular, A. L. , Bhappu, R. B. (Eds.) , Mineral Processing Plant Design, second ed. SME, New York, NY, USA, 679-702. (Chapter 33).

McGuire, P. M. , 2009. Conveyors: Application, Selection, and Integration. Industrial Innovation Series. Taylor & Francis Group, CRC Press, UK.

O'Brien, M. M. , Banks, H. R. , 1926. The Sullivan mine and concentrator. CIM Bull. 126, 1214-1235.

Patel, J. N. , et al. , 2013. Productivity improvement of screw conveyor by modified design. Int. J. Emerging Tech. Adv. Eng. 3 (1), 492-496.

Payant, R. , et al. , 2012. Galvanic interaction in self-heating of sulphide mixtures. CIM J. 3 (3), 169-177.

Pearse, G. , 1985. Pumps for the minerals industry. Mining Mag. Apr, 299-313.

Perry, R. H. , Green, D. W. , 1997. Perry's Chemical Engineers' Handbook. Seventh ed. McGraw-Hill, New York, NY, USA.

Quintiere, J. G. , et al. (2012). Spontaneous ignition in fire investigation. U. S. Department of Justice, National Critical Justice Reference Service Library, NCJ Document No: 239046.

Ray, S. , 2008. Introduction to Materials Handling. New Age International (P) Ltd. , New Delhi, India.

Roberts, A. W. , 2001. Recent developments in feeder design and performance. In: Levy, A. , Kalman, H. (Eds.) , Handbook of Conveying and Handling of Particulate Solids, Vol. 10. Elsevier, Amsterdam, New York, NY, USA, 211-223.

Rosenblum, F. , et al. , 2001. Evaluation and control of self-heating in sulphide concentrates. CIM Bull. 94 (1056), 92-99.

Rosenblum, F. , et al. , 2014. Review of self-heating testing methodologies. Proceedings of the 46th Annual Meeting Canadian Mineral Processors Conference, CIM, Ottawa, Canada, 67-89.

Somot, S. , Finch, J. A. , 2010. Possible role of hydrogen sulphide gas in self-heating of pyrrhotite-rich materials. Miner. Eng. 23 (2), 104-110.

Stracher, G. B. , 2007. Geology of Coal Fires: Case Studies from Around the World. Geological Society of America, Boulder, CO, USA, 279.

Walker, S. C. , 2012. Mine Winding and Transport. Elsevier, New York, NY, USA.

Wilson, G. , 1981. Selecting centrifugal slurry pumps to resist abrasive wear. Mining Eng. , SME, USA. 33, 1323-1327.

Wilson, K. C. , et al. , 2006. Slurry Transport Using Centrifugal Pumps. Third ed. Springer, New York, NY, USA.

Chapter 3 Classification

3.1 INTRODUCTION

Classification, as defined by Heiskanen (1993), is a method of separating mixtures of minerals into two or more products on the basis of the velocity with which the particles fall through a fluid medium. The carrying fluid can be a liquid or a gas. In mineral processing, this fluid is usually water, and wet classification is generally applied to mineral particles that are considered too fine ($<200\mu m$) to be sorted efficiently by screening. As such, this chapter will only discuss wet classification. A description of the historical development of both wet and dry classification is given by Lynch and Rowland (2005).

Classifiers are nearly always used to close the final stage of grinding and so strongly influence the performance of these circuits. Since the velocity of particles in a fluid medium is dependent not only on the size, but also on the specific gravity and shape of the particles, the principles of classification are also important in mineral separations utilizing gravity concentrators (Chapter 4).

3.2 PRINCIPLES OF CLASSIFICATION

3.2.1 Force Balance

When a solid particle falls freely in a vacuum, there is no resistance to the particle's motion. Therefore, if it is subjected to a constant acceleration, such as gravity, its velocity increases indefinitely, independent of size and density. Thus, a lump of lead and a feather fall at exactly the same rate in a vacuum.

In a viscous medium, such as air or water, there is resistance to this movement and this resistance increases with velocity. When equilibrium is reached between the gravitational force and the resistant force from the fluid, the body reaches its terminal velocity and thereafter falls at a uniform rate.

The nature of the resistance, or drag force, depends on the velocity of the descent. At low velocities, motion is smooth because the layer of fluid in contact with the body moves with it, while the fluid a short distance away is motionless. Between these two positions is a zone of intense shear in the fluid all around the descending particle. Effectively, all resistance to motion is due to the shear forces, or the viscosity of the fluid, and is hence called viscous resistance. At high velocities the main resistance is due to the displacement of fluid by the body, with the viscous resistance being relatively small; this is known as turbulent resistance. Whether viscous or turbulent resistance dominates, the acceleration of particles in a fluid rapidly decreases and the terminal velocity is reached relatively quickly.

A particle accelerates according to Newton's well known equation where $\sum F$ is the net force acting on a particle, m the mass of the particle, and a the acceleration of the particle:

$$\sum F = ma \qquad (3.1)$$

As mass, a combination of a particle's size and density, is a factor on the particle acceleration, it is common for fine high-density material, such as gold or galena, to be misclassified and report to the same product with the coarser low density particles. This occurrence will be further discussed in Section 3.2.4.

The classification process involves the balancing of the accelerating (gravitational, centrifugal, etc.) and opposing (drag, etc.) forces acting upon particles, so that the resulting net force has a different direction for fine and coarse particles. Classifiers are designed and operated so that the absolute velocities, resulting from the total net force, cause particles to be carried into separable products. Forces acting upon particles can include:

(1) Gravitational or electrostatic field force.

(2) Inertial force, centrifugal force, and Coriolis force (only in rotational systems).

(3) Drag force.

(4) Pressure gradient force, buoyancy force.

(5) Basset history force.

(6) Particle-particle interaction force.

An example of a classifier is a sorting column, in which a fluid is rising at a uniform rate (Fig. 3.1). Particles introduced into the sorting column either sink or rise according to whether their terminal velocities, a result of the net force, are greater or smaller than the upward velocity of the fluid. The sorting column therefore separates the feed into two products—an overflow consisting of particles with terminal velocities smaller than the velocity of the fluid and an underflow or spigot product containing particles with terminal velocities greater than the rising velocity.

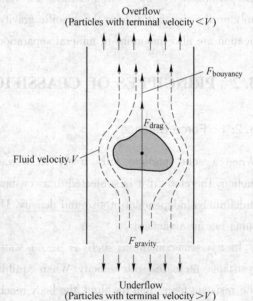

Fig. 3.1 Balance of forces on a particle in a sorting column

3.2.2 Free Settling

Free settling refers to the sinking of particles in a volume of fluid which is large with respect to the total volume of particles, hence particle-particle contact is negligible. For well dispersed pulps, free settling dominates when the percentage by weight of solids is less than about 15% (Taggart, 1945).

Consider a spherical particle of diameter d and density ρ_s falling under gravity in a viscous fluid of density ρ_f under free-settling conditions, that is, ideally in a fluid of infinite size. The particle is acted upon by three forces: a gravitational force acting downward (taken as positive direction), an upward buoyant force due to the displaced fluid, and a drag force D acting upward (see Fig. 3.1). Following Newton's law of motion in Eq. (3.1), the equation of motion of the particle is therefore:

$$\sum F = ma$$

$$F_{\text{Gravity}} - F_{\text{Buoyancy}} - F_{\text{Drag}} = m \frac{\mathrm{d}x}{\mathrm{d}t} \qquad (3.2)$$

$$mg - m'g - D = m \frac{\mathrm{d}x}{\mathrm{d}t}$$

where m is the mass of the particle, m' the mass of the displaced fluid, x the particle velocity, and g the acceleration due to gravity. When terminal velocity is reached, acceleration ($\mathrm{d}x/\mathrm{d}t$) is equal to zero, and hence:

$$D = g(m - m') \qquad (3.3)$$

Therefore, using the volume and density of a sphere:

$$D = \frac{\pi}{6} g d^3 (\rho_s - \rho_f) \qquad (3.4)$$

Stokes (1891) assumed that the drag force on a spherical particle was entirely due to viscous resistance and deduced the expression:

$$D = 3\pi d \eta v \qquad (3.5)$$

where η is the fluid viscosity and v is the terminal velocity.

Hence, substituting in Eq. (3.4) we derive:

$$3\pi d \eta v = \frac{\pi}{6} g d^3 (\rho_s - \rho_f)$$

or solving for the terminal velocity:

$$v = \frac{g d^2 (\rho_s - \rho_f)}{18 \eta} \qquad (3.6)$$

This expression is known as Stokes' law.

Newton assumed that the drag force was entirely due to turbulent resistance, and deduced:

$$D = 0.055 \pi d^2 v^2 \rho_f \qquad (3.7)$$

Substituting in Eq. (3.4) gives:

$$v = \left[\frac{3 g d (\rho_s - \rho_f)}{\rho_f} \right]^{1/2} \qquad (3.8)$$

This is Newton's law for turbulent resistance.

The range for which Stokes' law and Newton's law are valid is determined by the dimensionless

Reynolds number:

$$Re = \frac{vd\rho_f}{\eta}$$

where v is the terminal velocity of the particle (m/s), d the particle diameter (m), ρ_f the fluid density (kg/m^3), and η the fluid viscosity (Ns/m^2) ($\eta = 0.001$ Ns/m^2 for water at 20℃).

For Reynolds numbers below 1, Stokes'law is applicable. This represents, for quartz particles settling in water, particles below about 60μm in diameter. For higher Reynolds numbers, over 1000, Newton's law should be used (particles larger than about 0.5cm in diameter). There is, therefore, an intermediate range of Reynolds numbers (and particle sizes), which corresponds to the range in which most wet classification is performed, in which neither law fits experimental data. In this range there are a number of empirical equations that can be used to estimate the terminal velocity, some of which can be found in Heiskanen (1993).

Stokes'law (Eq. (3.6)) for a particular fluid can be simplified to:

$$v = k_1 d^2 (\rho_s - \rho_f) \tag{3.9}$$

And Newton's law (Eq. (3.8)) can be simplified to:

$$v = k_2 [d(\rho_s - \rho_f)]^{1/2} \tag{3.10}$$

where k_1 and k_2 are constants, and $(\rho_s - \rho_f)$ is known as the effective density of a particle of density ρ_s in a fluid of density ρ_f.

3.2.3 Hindered Settling

As the proportion of solids in the pulp increases above 15%, which is common in almost all mineral classification units, the effect of particle particle contact becomes more apparent and the falling rate of the particles begins to decrease. The system begins to behave as a heavy liquid whose density is that of the pulp rather than that of the carrier liquid; hindered-settling conditions now prevail. Because of the high density and viscosity of the slurry through which a particle must fall, in a separation by hindered settling the resistance to fall is mainly due to the turbulence created (Swanson, 1989). The interactions between particles themselves, and with the fluid, are complex and cannot be easily modeled. However, a modified form of Newton's law can be used to determine the approximate falling rate of the particles, in which ρ_p is the pulp density:

$$v = k_2 [d(\rho_s - \rho_p)]^{1/2} \tag{3.11}$$

3.2.4 Effect of Density on Separation Efficiency

The aforementioned laws show that the terminal velocity of a particle in a particular fluid is a function of the particle size and density. It can be concluded that:

(1) If two particles have the same density, then the particle with the larger diameter has the higher terminal velocity.

(2) If two particles have the same diameter, then the heavier (higher density) particle has the higher terminal velocity.

As the feed to most industrial classification devices will contain particles with varying densities, particles will not be classified based on size alone. Consider two mineral particles of densities ρ_a and ρ_b and diameters d_a and d_b, respectively, falling in a fluid of density ρ_f at exactly the same settling rate (their terminal velocity). Hence, from Stokes'law (Eq. (3.9)), for fine particles:

$$d_a^2(\rho_a - \rho_f) = d_b^2(\rho_b - \rho_f) \quad \text{or} \quad \frac{d_a}{d_b} = \left(\frac{\rho_b - \rho_f}{\rho_a - \rho_f}\right)^{1/2} \quad (3.12)$$

This expression is known as the free-settling ratio of the two minerals, that is, the ratio of particle size required for the two minerals to fall at equal rates.

Similarly, from Newton's law (Eq. (3.8)), the freesettling ratio of large particles is:

$$\frac{d_a}{d_b} = \frac{\rho_b - \rho_f}{\rho_a - \rho_f} \quad (3.13)$$

The general expression for free-settling ratio can be deduced from Eqs. (3.12) and (3.13) as:

$$\frac{d_a}{d_b} = \left(\frac{\rho_b - \rho_f}{\rho_a - \rho_f}\right)^n \quad (3.14)$$

where $n = 0.5$ for small particles obeying Stokes'law and $n = 1$ for large particles obeying Newton's law. The value of n lies in the range $0.5 \sim 1$ for particles in the intermediate size range of $50\mu m$ to $0.5 cm$ (Example 3.1).

The result in Example 3.1 means that the density difference between the particles has a more pronounced effect on classification at coarser size ranges. This is important where gravity concentration is being utilized. Over-grinding of the ore must be avoided, such that particles are fed to the separator in as coarse a state as possible, so that a rapid separation can be made, exploiting the enhanced effect of specific gravity difference. Since the enhanced gravity effect also means that fine heavy minerals are more likely to be recycled and overground in conventional ball mill-classifier circuits, it is preferable where possible to use open circuit rod mills for the primary coarse grind feeding a gravity circuit.

When considering hindered settling, the lower the density of the particle, the greater is the effect of the reduction of the effective density $(\rho_s - \rho_f)$. This then leads to a greater reduction in falling velocity. Similarly, the larger the particle, the greater is the reduction in falling rate as the pulp density increases. This is important in classifier design; in effect, hindered-settling reduces the effect of size, while increasing the effect of density on classification. This is illustrated by considering a mixture of quartz and galena particles settling in a pulp of density 1.5. The hindered-settling ratio can be derived from Eq. (3.11) as:

$$\frac{d_a}{d_b} = \frac{\rho_b - \rho_p}{\rho_a - \rho_p} \quad (3.15)$$

Therefore, in this system:

$$\frac{d_a}{d_b} = \frac{7.5 - 1.5}{2.65 - 1.5} = 5.22$$

A particle of galena will thus fall in the pulp at the same rate as a particle of quartz, which has a diameter 5.22 times as large. This compares with the free-settling ratio, calculated as 3.94 for turbulent resistance (Example 3.1).

The hindered-settling ratio is always greater than the free-settling ratio, and the denser the pulp, the greater is the ratio of the diameter of equal settling particles. For quartz and galena, the greatest hindered-settling ratio that we can attain practically is about 7.5. Hindered-settling classifiers are used to increase the effect of density on the separation, whereas free-settling classifiers use relatively dilute suspensions to increase the effect of size on the separation (Fig. 3.2). Relatively dense slurries are fed to certain gravity concentrators, particularly those treating heavy alluvial sands. This allows high tonnages to be treated and enhances the effect of specific gravity difference on the separation. The efficiency of

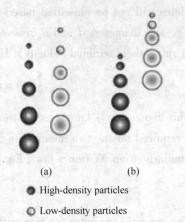

● High-density particles
○ Low-density particles

Fig. 3.2 Classification by
(a) free settling, and (b) hindered settling

separation, however, may be reduced since the viscosity of a slurry increases with density. For separations involving feeds with a high proportion of particles close to the required density of separation, lower slurry densities may be necessary, even though the density difference effect is reduced. As the pulp density increases, a point is reached where each mineral particle is covered only with a thin film of water. This condition is known as a quicksand, and because of surface tension, the mixture is a perfect suspension and does not tend to separate. The solids are in a condition of full teeter, which means that each grain is free to move, but is unable to do so without colliding with other grains and as a result stays in place. The mass acts as a viscous liquid and can be penetrated by solids with a higher specific gravity than that of the mass, which will then move at a velocity impeded by the viscosity of the mass.

A condition of teeter can be produced in a classifier sorting column by putting a constriction in the column, either by tapering the column or by inserting a grid into the base. Such hindered-settling sorting columns are known as teeter chambers. Due to the constriction, the velocity of the introduced water current is greatest at the bottom of the column. A particle falls until it reaches a point where its falling velocity equals that of the rising current. The particle can now fall no further. Many particles reach this condition, and as a result, a mass of particles becomes trapped above the constriction and pressure builds up in the mass. Particles move upward along the path of least resistance, which is usually the center of the column, until they reach a region of lower pressure at or near the top of the settled mass; here, under conditions in which they previously fell, they fall again. As particles from the bottom rise at the center, those from the sides fall into the resulting void. A general circulation is established, the particles being said to teeter. The constant jostling of teetering particles has a scouring effect which removes any entrained or adhering slimes particles, which then leave the teeter chamber and pass out through the classifier overflow. Clean separations can therefore be made in such classifiers. The teeter column principle is also exploited in some coarse particle flotation cells.

The analysis in this section has assumed, directly or implicitly, that the particles are spherical. While this is never really true, the resulting broken particle shapes are often close enough to spherical for the general findings above to apply. The obvious exceptions are ores containing flakey or fibrous mineral such as talc, mica, and some serpentines. Shape in those cases will influence the classification behavior.

Example 3.1

Consider a binary mixture of galena (specific gravity 7.5) and quartz (s. g. 2.65) particles being classified in water. Determine the free-settling ratio (a) from Stokes' law and (b) from Newton's law. What do you conclude?

Solution

(1) For small particles, obeying Stokes'law, the free-settling ratio (Eq. (3.12)) is:

$$\frac{d_a}{d_b} = \left(\frac{7.5 - 1}{2.65 - 1}\right)^{1/2} = 1.99$$

That is, a particle of galena will settle at the same rate as a particle of Quartz which has a diameter 1.99 times larger than the galena particle.

(2) For particles obeying Newton's law, the free-settling ratio (Eq. (3.13)) is:

$$\frac{d_a}{d_b} = \frac{7.5 - 1}{2.65 - 1} = 3.94$$

Conclusion: The free-settling ratio is larger for coarse particles obeying Newton's law than for fine particles obeying Stokes'law.

3.2.5 Effect of Classifier Operation on Grinding Circuit Behavior

Classifiers are widely employed in closed-circuit grinding operations to enhance the size reduction efficiency. The benefits of classification can include: improved comminution efficiency; improved product (classifier overflow) quality; and greater control of the circulating load to avoid overloading the circuit. The improvement in efficiency of the grinding circuit is seen as either a reduction in energy consumption or increase in throughput (capacity). The main increase in efficiency is due to the reduction of over grinding. By removing the finished product size particles from the circuit they are not subject to unnecessary further grinding (over grinding), which is a waste of comminution energy. This, combined with the recycling of unfinished (oversize) particles, results in the circuit product (classifier overflow) having a narrower size distribution than is the case for open circuit grinding. This narrow size distribution and restricted amount of excessively fine material benefits downstream mineral separation processes. Other benefits come from reduced particleparticle contact cushioning from fines in the grinding mill and less misplaced coarse material in the overflow, which would reduce downstream efficiencies. Therefore, classifier performance is critical to the optimal running of a mineral processing plant.

3.3 TYPES OF CLASSIFIERS

Although they can be categorized by many features, the most important is the force field applied to the unit: either gravitational or centrifugal. Centrifugal classifiers have gained widespread use as classifying equipment for many different types of ore. In comparison, gravitational classifiers, due to their low efficiencies at small particle sizes (<70μm), have limited use as classifiers and are only found in older plants or in some specialized cases. Table 3.1 outlines the key differences between the two types of classifiers and further highlights the benefits of centrifugal classifiers. Many types of classifiers have been designed and built, and only some common ones will be introduced. A more comprehensive guide to the major types of classification equipment used in mineral processing can be found elsewhere (Anon., 1984; Heiskanen, 1993; Lynch and Rowland, 2005).

Table 3.1 A Comparison of Key Parameters for Centrifugal and Gravitational Classifiers

Item	Centrifugal Classifiers	Gravitational Classifiers
Capacity	High	Low
Cut-size	Fine-Coarse	Coarse
Capacity/cut-size dependency	Yes	No
Energy consumption	High (feed pressure)	Low
Initial investment	Low	High
Footprint	Small	Large

3.4 CENTRIFUGAL CLASSIFIERS-THE HYDROCYCLONE

The hydrocyclone, commonly abbreviated to just cyclone, is a continuously operating classifying device that utilizes centrifugal force to accelerate the settling rate of particles. It is one of the most important devices in the minerals industry, its main use in mineral processing being as a classifier, which has proved extremely efficient at fine separation sizes. Apart from their use in closed-circuit grinding, cyclones have found many other uses, such as desliming, degritting, and thickening (dewatering). The reasons for this include their simplicity, low investment cost, versatility, and high capacity relative to unit size.

Hydrocyclones are available in a wide range of sizes, depending on the application, varying from 2.5m in diameter down to 10mm. This corresponds to cut-sizes of 300μm down to 1.5μm, with feed pressures varying from 20 to about 200kPa (3 ~ 30psi). The respective flowrates vary from $2m^3/s$ in a large unit to $2.5 \times 10^{-5} m^3/s$ in a small unit. Units can be installed on simple supports as single units or in clusters ("cyclopacs") (Heiskanen, 1993).

3.4.1 Basic Design and Operation

A typical hydro cyclone (Fig. 3.3) consists of a conically shaped vessel, open at its apex (also known as spigot or underflow), joined to a cylindrical section, which has a tangential feed inlet. The top of the cylindrical section is closed with a plate through which passes an axially mounted overflow pipe. The pipe is extended into the body of the cyclone by a short, removable section known as the vortex finder. The vortex finder forces the feed to travel downward, which prevents short-circuiting of feed directly into the overflow. The impact of these design parameters on performance is discussed further in Section 3.4.5.

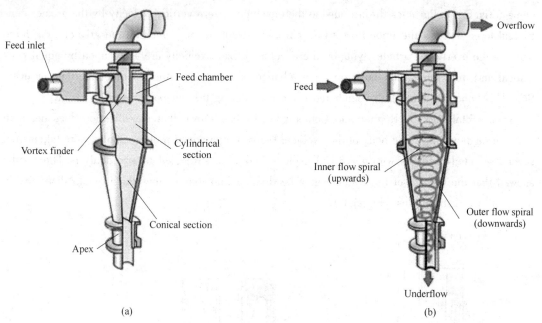

Fig. 3.3 A hydrocyclone showing
(a) main components, and (b) principal flows
(Adapted from Napier-Munn et al. (1996); courtesy JKMRC, The University of Queensland)

The feed is introduced under pressure through the tangential entry, which imparts a swirling motion to the pulp. This generates a vortex in the cyclone, with a low pressure zone along the vertical axis. An air core develops along the axis, normally connected to the atmosphere through the apex opening, but in part created by dissolved air coming out of solution in the zone of low pressure.

The conventional understanding is that particles within the hydro cyclone's flow pattern are subjected to two opposing forces: an outward acting centrifugal force and an inwardly acting drag (Fig. 3.4). The centrifugal force developed accelerates the settling rate of the particles, thereby separating particles according to size and specific gravity (and shape). Faster settling particles move to the wall of the cyclone, where the velocity is lowest, and migrate down to the apex opening. Due to the action of the drag force, the slower-settling particles move toward the zone of low pressure along the axis and are carried upward through the vortex finder to the overflow.

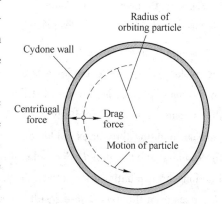

Fig. 3.4 Forces acting on an orbiting particle in the hydrocyclone

The existence of an outer region of downward flow and an inner region of upward flow implies a position at which there is no vertical velocity. This applies throughout the greater part of the cyclone body, and an envelope of zero vertical velocity should exist throughout the body of the cyclone (Fig. 3.5). Particles thrown outside the envelope of zero vertical velocity by the greater centrifugal force exit via the underflow, while particles swept to the center by the greater drag force leave in the overflow. Particles lying on the envelope of zero velocity are acted upon by equal centrifugal and drag forces and have an equal chance of reporting either to the underflow or overflow. This concept should be remembered when considering the cut-point described later.

Experimental work by Renner and Cohen (1978) has shown that classification does not take place throughout the whole body of the cyclone. Using a high-speed probe, samples were taken from several selected positions within a $\phi 150$ cyclone and were subjected to size analysis. The results showed that the interior of the cyclone may be divided into four regions that contain distinctively different size distributions (Fig. 3.6).

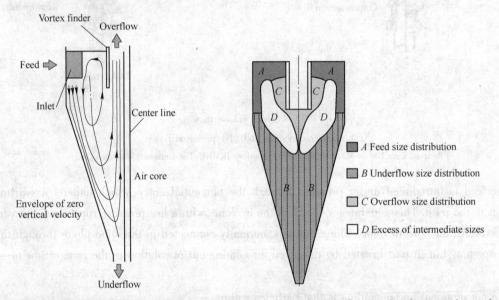

Fig. 3.5 Distribution of the vertical components of velocity in a hydrocyclone

Fig. 3.6 Regions of similar size distribution within cyclone (Renner and Cohen, 1978)

Essentially unclassified feed exists in a narrow region, A, adjacent to the cylinder wall and roof of the cyclone. Region B occupies a large part of the cone of the cyclone and contains fully classified coarse material, that is, the size distribution is practically uniform and resembles that of the underflow (coarse) product. Similarly, fully classified fine material is contained in region C, a narrow region surrounding the vortex finder and extending below the latter along the cyclone axis. Only in the toroid-shaped region D does classification appear to be taking place. Across this region, size fractions are radially distributed, so that decreasing sizes show maxima at decreasing radial distances from the axis. These results, however, were taken with a cyclone running at low pressure, so the region D may be larger in production units.

3.4.2 Characterization of Cyclone Efficiency

3.4.2.1 The Partition Curve

The most common method of representing classifier efficiency is by a partition curve (also known as a performance, efficiency, or selectivity curve). The curve relates the weight fraction of each particle size in the feed which reports to the underflow to the particle size. It is the same as the partition curve introduced for industrial screening. The cut-point, or separation size, is defined as the size for which 50% of the particles in the feed report to the underflow, that is, particles of this size have an equal chance of going either with the overflow or underflow (Svarovsky and Thew, 1992). This point is usually referred to as the d_{50} size.

It is observed in constructing the partition curve for a cyclone that the partition value does not appear to approach zero as particle size reduces, but rather approaches some constant value. To explain, Kelsall (1953) suggested that solids of all sizes are entrained in the coarse product liquid, bypassing classification in direct proportion to the fraction of feed water reporting to the underflow. For example, if the feed contains 16t/h of material of a certain size, and 12t/h reports to the underflow, then the percentage of this size reporting to the underflow, and plotted on the partition curve, is 75%. However, if, say, 25% of the feed water reports to the underflow, then 25% of the feed material will short-circuit with it; therefore, 4t/h of the size fraction will short-circuit to the underflow, and only 8t/h leave in the underflow due to classification. The corrected recovery of the size fraction is thus:

$$100 \times \frac{12 - 4}{16 - 4} = 67\%$$

The uncorrected, or actual, partition curve can therefore be corrected by utilizing Eq. (3.16):

$$C = \frac{S - R_{w/u}}{1 - R_{w/u}} \qquad (3.16)$$

where C is the corrected mass fraction of a particular size reporting to underflow, S the actual mass fraction of a particular size reporting to the underflow, and $R_{w/u}$ the fraction of the feed liquid that is recovered in the coarse product (underflow) stream, which defines the bypass fraction. The corrected curve thus describes particles recovered to the underflow by true classification and introduces a corrected cut-size, d_{50c}. Kelsall's assumption has been questioned, and Flintoff et al. (1987) reviewed some of the arguments. However, the Kelsall correction has the advantages of simplicity, utility, and familiarity through long use.

The construction of the actual and corrected partition curves can be illustrated by means of an example. The calculations are easily performed in a spreadsheet (Example 3.2).

Modern literature, including manufacturer's data, usually quotes the corrected d_{50} with the subscript "c" dropped. Some care in reading the literature is therefore required. To avoid confusion, the designation d_{50c} will be retained.

Although not demonstrated here, the partition curve is often plotted against d/d_{50}, that is, making the size axis nondimensional—a reduced or normalized size—giving rise to a reduced partition curve.

The exponential form of the curves has led to several fitting models (Napier-Munn et al., 1996). One of the most common (Plitt, 1976) is:

$$C = 1 - \exp(-Kd^m) \qquad (3.17)$$

here K is a constant, d the mean particle size, and m the "sharpness" of separation. By introducing $C = 0.5$ at $d = d_{50c}$ and rearranging Eq. (3.17) gives K:

$$K = \frac{\ln 0.5}{d_{50c}^m} = \frac{0.693}{d_{50c}^m}$$

thus
$$C = 1 - \exp\left[-0.693\left(\frac{d}{d_{50c}}\right)^m\right] \qquad (3.18a)$$

and
$$S = C(1 - R_{w/u}) + R_{w/u} \qquad (3.18b)$$

The normalized size is evident in Eq. (3.18a). By fitting data to Eq. (3.18a), the two unknowns, d_{50c} and m, can be estimated.

Example 3.2

(1) Determine the actual partition curve and actual d_{50} for the cyclone data.

(2) Determine the corrected partition curve and corrected d_{50c}.

As a reminder, the grinding circuit is

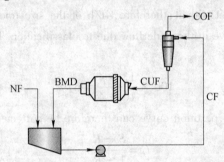

Solution

(1) Using the symbolism then the partition value S is given by:

$$S = \frac{U_u}{F_f}$$

From the size distribution data using the generalized least squares minimization procedure, the solids split (recovery) to underflow (U/F) is 0.619. From this, data reconciliation was executed and the adjusted size distribution data are given in Table EX 3.2 along with the actual partition (S) and the corrected partition (C) values.

Specimen calculation of S and mean particle size:

For the $-592+419\mu m$ size class:

$$S = 0.619 \times \frac{7.51}{4.86} = 0.957 \text{ or } 95.7\%$$

The mean size is calculated by taking the geometric mean; for example, for the $-592+419$ size class this is $(592 \times 419)^{0.5} = 498\mu m$.

3.4 CENTRIFUGAL CLASSIFIERS-THE HYDROCYCLONE

The partition curve is then constructed by plotting the S values (as % in this case) versus the mean particle size, as done in Fig. EX 3.2-S.

From the actual partition curve the cut-point d_{50} is about 90μm.

(2) It showed that the water split (recovery) to underflow was 0.32 (32%). From the actual partition curve in Fig. EX 3.2-S this appears to be a reasonable estimate of the bypass fraction. Using this bypass value the corrected partition C values are computed and the corrected partition curve constructed which is included in the figure.

Specimen calculation:

For the $-209+148$ μm size class:

$$C = \frac{0.743 - 0.32}{1 - 0.32} = 0.622 \text{ or } 62.2\%$$

From the corrected partition curve the d_{50c} is about 120μm.

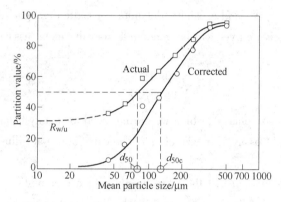

Fig. EX 3.2-S Actual and corrected Partition curves for data in the table

Table EX 3.2-S Adjusted Size Distribution Data and Partition Values

Size interval/μm	Mean size/μm	CF	COF	CUF	Partition, S	Corrected, C
+592		8.27	0.23	13.21	98.9	98.3
−592+419	498	4.86	0.54	7.51	95.7	93.6
−419+296	352	6.77	0.94	10.35	94.6	92.1
−296+209	249	7.04	2.81	9.64	84.8	77.6
−209+148	176	11.38	7.67	13.66	74.3	62.2
−148+105	125	13.91	13.21	14.35	63.9	46.9
−105+74	88	11.09	11.7	10.71	59.8	40.9
−74+53	63	9.86	14.78	6.83	42.9	16.0
−53+37	44	5.57	9.33	3.26	36.2	6.2
−37		21.25	38.79	10.48	30.5	−2.2
		100	100	100		

3.4.2.2 Sharpness of Cut

The value of m in Eq. (3.18a) is one measure of the sharpness of the separation. The value is determined by the slope of the central section of the partition curve; the closer the slope is to vertical, the higher the value of m and the greater the classification efficiency. Perfect classification would give $m = \infty$; but in reality m values are rarely above 3. The slope of the curve can be also expressed by taking the points at which 75% and 25% of the feed particles report to the underflow. These are the d_{75} and d_{25} sizes, respectively. The sharpness of separation, or the so-called imperfection, I, is then given by (where the actual or corrected d_{50} may be used):

$$I = \frac{d_{75} - d_{25}}{2d_{50}} \tag{3.19}$$

3.4.2.3 Multidensity Feeds

Inspection of the partition curve in Example 3.2 (Fig. EX 3.2-S) shows a deviation at about 100μm which we chose to ignore by passing a smooth curve through the data, implying, perhaps, some experimental uncertainty. In fact, the deviation is real and is created by classifying feeds with minerals of different specific gravity. In this case the feed is a Pb-Zn ore with galena (s.g. 7.5), sphalerite (4.0), pyrite (5.0), and a calcite/dolomite non-sulfide gangue, NSG (2.85). The individual mineral classification curves can be generated from the metal assay on each size fraction using the following:

$$S_{i,M} = \frac{Uu_i u_{i,M}}{Ff f_{i,M}} \tag{3.20}$$

where subscript i refers to size class i (which we did not need to specify before) and subscript i,M refers to metal (or mineral) assay M in size class i. The calculations are illustrated using an example (Example 3.3).

From Example 3.3 it is evident that the high-density galena preferentially reports to the underflow. The minerals, in fact, classify according to their density. Treating the elemental assay curves to be analogous with their associated minerals has the implication that they are liberated, free minerals. This is not entirely true, after all, the purpose of the grinding circuit is to liberate, so in the cyclone feed liberation is unlikely to be complete. It is better to refer to mineral-by-size curves, which does not imply liberated minerals. The analysis can be extended to determine the corrected partition curves and the corresponding m and d_{50c}, M values, which we can then use as a first approximation for the associated minerals. In doing so we find the following values: $m = 2.2$ for all minerals, and d_{50c}, M equals 37μm for galena, 66μm for pyrite; 89μm for sphalerite, and 201μm for NSG. The m value shows the sharpness of separation is higher for the minerals than for the total solids ($m = 1.7$, Example 3.2). By inspection, the d_{50c}, M values correspond well with the hindered settling expression, from Eq. (3.15). For example, the ratio d_{50c}, NSG: d_{50c}, Ga is 5.4, which is close to the value calculated for the hindered settling of essentially this pair of minerals. It is this density effect, concentrating heavy minerals in the circulating stream, which gives the opportunity for recovery from this stream (or other streams inside the circuit), using gravity separation devices or flash flotation.

Example 3.3

(1) Determine the actual partition values for Pb (galena) from the data in the table.

(2) Determine the circulating load of galena and compare with the circulating load of solids.

Solution

(1) The table and the calculation of the partition values for galena (Si, Pb) calculated using Eq. (3.20) are shown in Table EX 3.3-S.

Table EX 3.3-S Size Distribution and Pb Assay by Size for the Same Circuit as in Example 3.2-S

Size interval/μm	Mean size/μm	CF		COF		CUF		$S_{i,Pb}$
		w/%	w(Pb)/%	w/%	w(Pb)/%	w/%	w(Pb)/%	
+592		8.27	1.33	0.23	1.41	13.21	1.32	98.1
−592+419	498	4.86	1.02	0.54	1.3	7.51	1	93.8
−419+296	352	6.77	1.73	0.94	−0.16	10.35	1.84	100.7
−296+209	249	7.04	2.55	2.81	−0.12	9.64	3.02	100.4
−209+148	176	11.38	3.19	7.67	0.25	13.66	4.21	98.1
−148+105	125	13.91	3.83	13.21	0.22	14.35	5.87	97.9
−105+74	88	11.09	8.14	11.7	0.35	10.71	13.36	98.1
−74+53	63	9.86	9.21	14.78	1.02	6.83	20.07	93.4
−53+37	44	5.57	6.77	9.33	2.29	3.26	14.64	78.3
−37		21.25	3.55	38.79	3.09	10.48	4.59	39.5
Head		100	4.29	100	1.66	100	5.91	85.3

Note: The "head" row gives the total Pb assay for the stream; and that data reconciliation was not constrained to zero on the Pb assay giving a couple of small negative values in the COF column and consequently $S_{i,Pb}$ values slightly larger than 100%.

Specimen calculation:

For the −53+37μm size class:

$$S_{i,Pb} = 0.619 \times \frac{3.26 \times 14.64}{5.57 \times 6.77} = 0.783 \text{ or } 78.3\%$$

The actual partition curves for all minerals are given in Fig. EX 3.3-S.

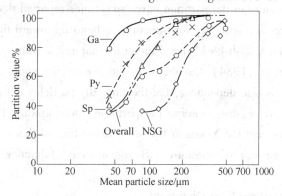

Fig. EX 3.3-S Actual partition curves for minerals and total solids (overall)

(2) The total solids circulating load CLTot and total galena circulating load CLTot,Pb are given by:

$$CL_{Tot} = \frac{U}{O} = \frac{0.619}{0.381} = 1.63$$

$$CL_{Tot,Pb} = \frac{Uu_{Tot,Pb}}{Oo_{Tot,Pb}} = 1.63 \times \frac{5.91}{1.66} = 5.78$$

The circulating load of galena is thus 3.56 (5.78/1.63) times that of the solids.

3.4.2.4 Unusual Partition Curves

"Unusual" refers to deviations from the simple exponential form, which can range from inflections to the curve displaying a minimum and maximum (Laplante and Finch, 1984; Kawatra and Eisele, 2006). The partition curve in Example 3.2 illustrates the phenomena: the inflection at about 100μm was initially ignored in favor of a smooth curve, suggesting, perhaps, experimental error. Example 3.3 shows that the inflection in the total solids curve is because it is the sum of the mineral partition curves (weighted for the mineral content); below 100μm the NSG is no longer being classified and the solids curve shifts over to reflect the continuing classification of the denser minerals. The "unusual" nature of the curve is even more evident in series cyclones where the downstream (secondary) cyclone receives overflow from the primary cyclone, which now comprises fine dense mineral particles and coarse light mineral particles.

The impact of the component minerals is also evident in the m values: that for the mineral curves is higher than for the solids, and the correspondence to the exponential model, Eq. (3.18a), is better for the minerals than for the solids. It should be remembered when determining m that a low value may not indicate performance that needs improving, as the component minerals may be comparatively quite sharply classified.

Apart from density, the shape of the particles in the feed is also a factor in separation. Flakey particles such as mica often reporting to the overflow, even though they may be relatively coarse. Wet classification of asbestos feeds can reveal similar unusual overall partition curves as for density, this time the result of the shape: fibrous asbestos minerals, and the granular rock particles.

Another "unusual" feature of the partition curve sometimes reported that is not connected to a density (or shape) effect is a tendency for the curve to bend up toward the fine end (Del Villar and Finch, 1992). Called a "fish-hook" by virtue of its shape, it is well recognized in mechanical air classifiers (Austin et al., 1984). Connected to the bypass, the notion is that the fraction reporting to the underflow is size dependent, that the finer the particles the closer they split in the same proportion as the water, thus bending the partition curve upward (the fish-hook) at the finer particle sizes to intercept the Y-axis at $R_{w/u}$. Whether the fish-hook is real or not continues to attract a surprising amount of literature (Bourgeois and Majumder, 2013; Nageswararao, 2014).

3.4.2.5 Cyclone Overflow Size Distribution

Although partition curves are useful in assessing and modeling classifier performance, the minerals engineer is usually more interested in knowing fineness of grind (i.e., cyclone overflow particle size distribution) than the cyclone cut-size. Simple relationships between fineness of grind and the partition curve of a hydrocyclone have been developed by Arterburn (1982) and Kawatra and Seitz (1985).

Fig. 3.7(a) shows the evolution of size distribution of the solids from the feed to product streams for the grinding circuit in Example 3.2; and Fig. 3.7(b) shows the size distribution of the solids and the minerals in the cyclone overflow for the same circuit. The latter shows the much finer size distribution of the galena ($P_{80,Pb} \sim 45\mu m$) compared to the solids ($P_{80} \sim 120\mu m$) and the NSG ($P_{80,NSG} \sim 130\mu m$) due to the high circulating load, and thus additional grinding, of the galena. It is this observation which gives rise to the argument that the grinding circuit treating a high-density mineral component can be operated at a coarse solids P_{80}, as the high-density mineral will be automatically ground finer, and the grind instead could be made the P_{80} of the target mineral for recovery.

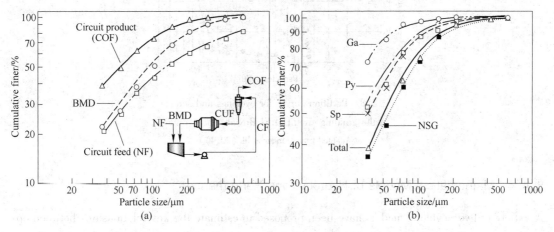

Fig. 3.7 (a) Evolution of size distribution through closed grinding circuit (Example 3.2), and (b) size distribution of minerals in the overflow product from the same circuit

This density effect in cyclones (or any classification device) is often taken to be a reason to consider screening which is density independent. This is a good point at which to compare cyclones with screens.

3.4.3 Hydrocyclones Versus Screens

Hydrocyclones have come to dominate classification when dealing with fine particle sizes in closed grinding circuits (<200μm). However, recent developments in screen technology have renewed interest in using screens in grinding circuits. Screens separate on the basis of size and are not directly influenced by the density spread in the feed minerals. This can be an advantage. Screens also do not have a bypass fraction, and as Example 3.2 has shown, bypass can be quite large (over 30% in that case). Fig. 3.8 shows an example of the difference in partition curve for cyclones and screens. The data is from the El Brocal concentrator in Peru with evaluations before and after the hydrocyclones were replaced with a Derrick Stack Sizers in the grinding circuit (Dundar et al., 2014). Consistent with expectation, compared to the cyclone the screen had a sharper separation (slope of curve is higher) and little bypass. An increase in grinding circuit capacity was reported due to higher breakage rates after implementing the screen. This was attributed to the elimination of the bypass, reducing the amount of fine material sent back to the grinding mills which tends to cushion particleparticle impacts. Changeover is not one way, however: a recent example is a switch from screen to cyclone, to take advantage of the additional size reduction of the denser payminerals (Sasseville, 2015).

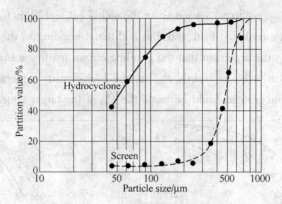

Fig. 3.8 Partition curves for cyclones and screens
in the grinding circuit at El Brocal concentrator
(Adapted from Dundar et al. (2014))

3.4.4 Mathematical Models of Hydrocyclone Performance

A variety of hydrocyclone models have been proposed to estimate the key relationships between operating and geometrical variables for use in design and optimization, with some success. These include empirical models calibrated against experimental data, as well as semi-empirical models based on equilibrium orbit theory, residence time, and turbulent flow theory. Progress is also being made in using computational fluid dynamics to model hydrocyclones from first principles (e.g., Brennan et al.,2003;Nowakowski et al.,2004;Narasimha et al.,2005). All the models commonly used in practice are still essentially empirical in nature.

3.4.4.1 Bradley model

Bradley's seminal book (1965) listed eight equations for the cut-size, and the number has increased significantly since then. Bradley's equation based on the equilibrium orbit hypothesis (Fig. 3.4) was:

$$d_{50} = k \left[\frac{D_c^3 \eta}{Q_f (\rho_s - \rho_1)} \right]^n \tag{3.21}$$

where D_c is the cyclone diameter, η the fluid viscosity, Q_f the feed flowrate, ρ_s the solids density, ρ_1 the fluid density, n a hydrodynamic constant (0.5 for particle laminar flow), and k a constant incorporating other factors, particularly cyclone geometry, which must be estimated from experimental data. Eq. (3.21) describes some of the process trends well, but cannot be used directly in design or operational situations.

3.4.4.2 Empirical Models

A variety of empirical models were constructed in the previous century to predict the performance of hydrocyclones (e.g.,Leith and Licht,1972). More recently, models have been proposed by, among others,Nageswararao et al. (2004) and Kraipech et al. (2006). The most widely used empirical models are probably those of Plitt (1976) and its later modified form (Flintoff et al., 1987),and Nageswararao (1995). These models, based on a phenomenological description of the process with numerical constants determined from large databases, are described in Napier-Munn et al. (1996) and were reviewed and compared by Nageswararao et al. (2004).

Plitt's modified model for the corrected cut-size d_{50c} in micrometers is:

$$d_{50c} = \frac{F_1 39.7 D_c^{0.46} D_i^{0.6} D_o^{1.21} \eta^{0.5} \exp(0.06 C_v)}{D_u^{0.71} h^{0.38} Q_f^{0.45} \left(\frac{\rho_s - 1}{1.6}\right)^k} \tag{3.22}$$

where D_c, D_i, D_o, and D_u are the inside diameters of the cyclone, inlet, vortex finder, and apex, respectively (cm), η the liquid viscosity (cP), C_v the feed solids volume concentration (%), h the distance between apex and end of vortex finder (cm), k a hydrodynamic exponent to be estimated from data (default value for laminar flow 0.5), Q_f the feed flowrate (1min^{-1}), and ρ_s the solids density (g/cm^3). Note that for noncircular inlets, $D_i = \sqrt{4A/\pi}$ where A is the cross-sectional area of the inlet (cm^2).

The equation for the volumetric flowrate of slurry to the cyclone Q_f is:

$$Q_f = \frac{F_2 P^{0.56} D_c^{0.21} D_i^{0.53} h^{0.16} (D_u^2 + D_o^2)^{0.49}}{\exp(0.0031 C_v)} \tag{3.23}$$

where P is the pressure drop across the cyclone in kilopascals (1psi = 6.895kPa). The F_1 and F_2 in Eqs. (3.22) and (3.23) are material-specific constants that must be determined from tests with the feed material concerned. Plitt also reports equations for the flow split between underflow and overflow, and for the sharpness of separation parameter m in the corrected partition curve.

Nageswararao's model includes correlations for corrected cut-size, pressure-flowrate, and flow split, though not sharpness of separation. It also requires the estimate of feed-specific constants from data, though first approximations can be obtained from libraries of previous case studies. This requirement for feed-specific calibration emphasizes the important effect that feed conditions have on hydrocyclone performance. Asomah and Napier-Munn (1997) reported an empirical model that incorporates the angle of inclination of the cyclone, as well as explicitly the slurry viscosity, but this has not yet been validated in the large-scale use which has been enjoyed by the Plitt and Nageswararao models.

A useful general approximation for the flowrate in a hydrocyclone is:

$$Q \approx 9.5 \times 10^{-3} D_c^2 \sqrt{P} \tag{3.24}$$

Flowrate and pressure drop together define the useful work done in the cyclone:

$$\text{Power} = \frac{PQ}{3600} (\text{kW}) \tag{3.25}$$

where Q is the flowrate (m^3/h); P the pressure drop (kPa); and D_c the cyclone diameter (cm). The power can be used as a first approximation to size the pump motor, making allowances for head losses and pump efficiency.

These models are easy to incorporate in spreadsheets, and so are particularly useful in process design and optimization using dedicated computer simulators such as JKSimMet (Napier-Munn et al. ,1996) and MODSIM (King,2012), or the flowsheet simulator Limn (Hand and Wiseman, 2010). They can also be used as a virtual instrument or "soft sensor" (Morrison and Freeman, 1990; Smith and Swartz, 1999), inferring cyclone product size from geometry and operating variables as an alternative to using an online particle size analyzer.

3.4.4.3 Computational Models

With advances in computer hardware and software, considerable progress has been made in the fundamental modeling of hydrocyclones. The multiphase flow within a hydrocyclone consists of solid particles, which are dispersed throughout the fluid, generally water. In addition, an air core is present. Such multiphase flows can be studied using a combination of Computational Fluid Dynamics (CFD) and Discrete Element Method (DEM) techniques.

The correct choice of the turbulence, multiphase (airwater interaction), particle drag, and contact models are essential for the successful modeling of the hydrocyclone. The highly turbulent swirling flows, along with the complexity of the air core and the relatively high feed percent solids, incur large computational effort. Choices are therefore made based on a combination of accuracy and computational expense. Studies have included effects of short-circuiting flow, motion of different particle sizes, the surging phenomena, and the "fish-hook" effect as well as particleparticle, particlefluid, and particlewall interactions in both dense medium cyclones (Chapter 5) and hydrocyclones.

The data required to validate numerical models can be obtained through methods such as particle image velocimetry and laser Doppler velocimetry. These track the average velocity distributions of particles, as opposed to Lagrangian tracking, where individual particles of a dispersed phase can be tracked in space and time. Wang et al. (2008) used a high-speed camera to record the motion of a particle with density $1140kg/m^3$ in water in a hydrocyclone. The two-dimensional particle paths obtained showed that local or instantaneous instabilities in the flow field can have a major effect on the particle trajectory, and hence on the separation performance of the hydrocyclone. Lagrangian tracking includes the use of positron emission particle tracking (PEPT), which is a more recent development in process engineering. PEPT locates a point-like positron emitter by cross-triangulation. Chang et al. (2011) studied the flow of a particle through a hydrocyclone using PEPT with an 18F radioactive tracer and could track the particle in the cyclone with an accuracy of 0.2mm/ms.

3.4.5 Operating and Geometric Factors Affecting Cyclone Performance

The empirical models and scale-up correlations, tempered by experience, are helpful in summarizing the effects of operating and design variables on cyclone performance. The following process trends generally hold true: Cut-size (inversely related to solids recovery):

(1) Increases with cyclone diameter.

(2) Increases with feed solids concentration and/or viscosity.

(3) Decreases with flowrate.

(4) Increases with small apex or large vortex finder.

(5) Increases with cyclone inclination to vertical Classification efficiency.

(6) Increases with correct cyclone size selection.
(7) Decreases with feed solids concentration and/or viscosity.
(8) Increases by limiting water to underflow.
(9) Increases with certain geometries Flow split of water to underflow.
(10) Increases with larger apex or smaller vortex finder.
(11) Decreases with flowrate.
(12) Decreases with inclined cyclones (especially low pressure).
(13) Increases with feed solids concentration and/or viscosity Flowrate.
(14) Increases with pressure.
(15) Increases with cyclone diameter.
(16) Decreases (at a given pressure) with feed solids concentration and/or viscosity.

Since the operating variables have an important effect on the cyclone performance, it is necessary to avoid fluctuations, such as in flowrate and pressure drop, during operation. Pump surging should be eliminated either by automatic control of level in the sump, or by a self regulating sump, and adequate surge capacity should be installed to eliminate flowrate fluctuations.

The feed flowrate and the pressure drop across the cyclone are related (Eq. (3.22)). The pressure drop is required to enable design of the pumping system for a given capacity or to determine the capacity for a given installation. Usually the pressure drop is determined from a feed-pressure gauge located on the inlet line some distance upstream from the cyclone. Within limits, an increase in feed flowrate will improve fine particle classification efficiency by increasing the centrifugal force on the particles. All other variables being constant, this can only be achieved by an increase in pressure and a corresponding increase in power, since this is directly related to the product of pressure drop and capacity. Since increase in feed rate, or pressure drop, increases the centrifugal force, finer particles are carried to the underflow, and d_{50} is decreased, but the change has to be large to have a significant effect. Fig. 3.9 shows the effect of pressure on the capacity and cut-point of cyclones.

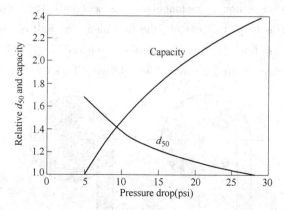

Fig. 3.9 Effect of pressure on capacity and cut-point of hydrocyclone

The effect of increase in feed pulp density is complex, as the effective pulp viscosity and degree of hindered settling is increased within the cyclone. The sharpness of the separation decreases with increasing pulp density and the cut-point rises due to the greater resistance to the swirling motion within the cyclone, which reduces the effective pressure drop. Separation at finer sizes can only be achieved with feeds of low solids content and large pressure drop. Normally, the feed concentration is no greater than about 30% solids by weight, but for closed-circuit grinding operations, where relatively coarse separations are often required, high feed concentrations of up to 60% solids by weight are often used, combined with low pressure drops, often less than 10psi (68.9kPa). Fig. 3.10 shows that feed concentration has an important effect on the cut-size at high pulp densities.

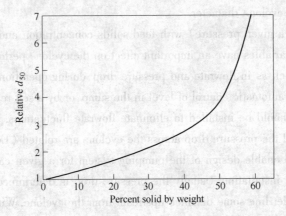

Fig. 3.10 Effect of solids concentration on cut-point of hydrocyclone

In practice, the cut-point is mainly controlled by the cyclone design variables, such as those of the inlet, vortex-finder, and apex openings, and most cyclones are designed such that these are easily changed.

The area of the inlet determines the entrance velocity and an increase in area increases the flow-rate. The geometry of the feed inlet is also important. In most cyclones the shape of the entry is developed from circular cross section to rectangular cross section at the entrance to the cylindrical section of the cyclone. This helps to "spread" the flow along the wall of the chamber. The inlet is normally tangential, but involuted feed entries are also common (Fig. 3.11). Involuted entries are said to minimize turbulence and reduce wear. Such design differences are reflected in proprietary

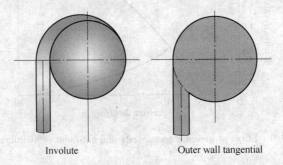

Fig. 3.11 Involute and tangential feed entries

cyclone developments such as Weir Warman's CAVEX® and Krebs'gMAX® units.

The diameter of the vortex finder is an important variable. At a given pressure drop across the cyclone, an increase in the diameter of the vortex finder will result in a coarser cut-point and an increase in capacity.

The size of the apex opening determines the underflow density and must be large enough to discharge the coarse solids that are being separated by the cyclone. The orifice must also permit the entry of air along the axis of the cyclone to establish the air vortex. Cyclones should be operated at the highest possible underflow density, since unclassified material (the bypass fraction) leaves the underflow in proportion to the fraction of feed water leaving via the underflow. Under correct operating conditions, the discharge should form a hollow cone spray with a 20 ~ 30 included angle (Fig. 3.12(a)). Air can then enter the cyclone, the classified coarse particles will discharge freely, and solids concentrations greater than 50% by weight can be achieved. Too small apex opening can lead to the condition known as "roping" (Fig. 3.12(b)), where an extremely thick pulp stream of the same diameter as the apex is formed, and the air vortex may be lost, the separation efficiency will fall, and oversize material will discharge through the vortex finder (This condition is sometimes encouraged where a very high underflow solids concentration is required, but is otherwise deleterious: the impact of this "tramp" oversize on downstream flotation operations can be quite dramatic (Bahar et al., 2014)). Too large an apex orifice results in the larger hollow cone pattern seen in Fig. 3.12(c). The underflow will be excessively dilute and the additional water will carry unclassified fine solids that would otherwise report to the overflow. The state of operation of a cyclone is important to optimize both grinding efficiency and downstream separation processes, and online sensors are now being incorporated to identify malfunctioning cyclones (Cirulis and Russell, 2011; Westendorf et al., 2015).

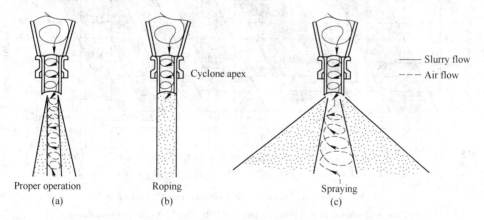

Fig. 3.12 Nature of underflow discharge

(a) correct apex size-proper operation, (b) apex too small-"roping" leading to loss of air core, and (c) apex too large-"spraying" which leads to lower sharpness of separation

Some investigators have concluded that the cyclone diameter has no effect on the cut-point and that for geometrically similar cyclones the efficiency curve is a function only of the feed material characteristics. The inlet and outlet diameters are the critical design variables, the cyclone diameter merely being the size of the housing required to accommodate these apertures (Lynch et al. ,1974, 1975; Rao et al. ,1976). This is true where the inlet and outlet diameters are essentially proxies for cyclone diameter for geometrically similar cyclones. However, from theoretical considerations, it is the cyclone diameter that controls the radius of orbit and thus the centrifugal force acting on the particles. As there is a strong interdependence between the aperture sizes and cyclone diameter, it is difficult to distinguish the true effect, and Plitt (1976) concluded that the cyclone diameter has an independent effect on separation size.

For geometrically similar cyclones at constant flowrate, $d_{50} \propto \text{diameter}^x$, but the value of x is open to much debate. The value of x using the KrebsMularJull model is 1.875, for Plitt's model it is 1.18, and Bradley (1965) concluded that x varies from 1.36 to 1.52.

3.4.6 Sizing and Scale-Up of Hydrocyclones

In practice, the cut-point is determined to a large extent by the cyclone size (diameter of cylindrical section). The size required for a particular application can be estimated from empirical models (discussed below), but these tend to become unreliable with extremely large cyclones due to the increased turbulence within the unit, and it is therefore more common to choose the required model by referring to manufacturers' charts, which show capacity and separation size range in terms of cyclone size. A typical performance chart is shown in Fig. 3.13 (where the D designation refers to cyclone diameter in inches). This is for Krebs cyclones, operating at less than 30% feed solids by weight, and with solids specific gravity in the range 2.5~3.2.

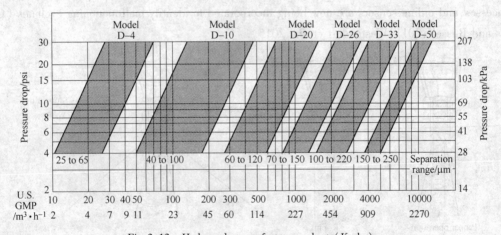

Fig. 3.13 Hydrocyclone performance chart (Krebs)

(From Napier-Munn et al. (1996); courtesy JKMRC, The University of Queensland)

Since fine separations require small cyclones, which have only small capacity, enough have to be connected in parallel if to meet the capacity required; these are referred to as clusters, batteries, nests, or cyclopacs (Fig. 3.14). Cyclones used for desliming duties are usually very small in diameter, and a large number may be required if substantial flowrates must be handled. The desliming

Fig. 3.14 A nest of 150mm cyclones at the Century Zinc mine, Australia
(Courtesy JKMRC and JKTech Pty Ltd.)

plant at the Mr. Keith Nickel concentrator in Western Australia has 4000 such cyclones. The practical problems of distributing the feed evenly and minimizing blockages have been largely overcome by the use of Mozley cyclone assemblies. A 16×51mm (16×2in.) assembly is shown in Fig. 3.15. The feed is introduced into a housing via a central inlet at pressures of up to 50psi (344.8kPa). The housing contains 16×2in. cyclones, which have interchangeable spigot and vortex finder caps for precise control of cut-size and underflow density. The feed is forced through a trash screen and into each cyclone without the need for separate distributing ports (Fig. 3.16). The overflow from each cyclone leaves via the inner pressure plate and leaves the housing through the single overflow pipe on the side. The assembly design reduces maintenance, the removal of the top cover allowing easy access so that individual cyclones can be removed (for repair or replacement) without disconnecting feed or overflow pipework.

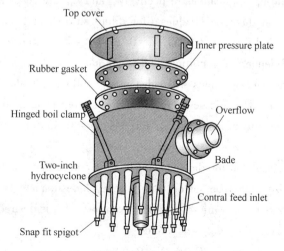

Fig. 3.15 Mozley 16×2in. cyclone assembly

Fig. 3.16 Interior of Mozley 16×2in. cyclone assembly

Since separation at large particle size requires large diameter cyclones with consequent high capacities, in some cases, where coarse separations are required, cyclones cannot be utilized, as the plant throughput is not high enough. This is often a problem in pilot plants where simulation of the full-size plant cannot be achieved, as scaling down the size of cyclones to allow for smaller capacity also reduces the cut-point produced.

(1) Scale-Up of Hydrocyclones. A preliminary scale-up from a known situation (e.g., a laboratory or pilot plant test) to the unknown (e.g., a full production installation) can be done via the basic relationships between cut-size, cyclone diameter, flowrate, and pressure drop. These are:

$$\frac{d_{50c2}}{d_{50c1}} = \left(\frac{D_{c2}}{D_{c1}}\right)^{n_1}\left(\frac{Q_1}{Q_2}\right)^{n_2} = \left(\frac{D_{c2}}{D_{c1}}\right)^{n_3}\left(\frac{P_1}{P_2}\right)^{n_4} \quad (3.26)$$

and

$$\frac{P_1}{P_2} \cong \left(\frac{Q_1}{Q_2}\right)^{n_5}\left(\frac{D_{c2}}{D_{c1}}\right)^{n_6} \quad (3.27)$$

where P is pressure drop, Q the flowrate, D_c the cyclone diameter, the subscripts 1 and 2 indicate the known and scale-up applications respectively, and n_{1-6} are constants which are a function of the flow conditions. The theoretical values (for dilute slurries and laminar flow in small cyclones) are: $n_1 = 1.5$, $n_2 = 0.5$, $n_3 = 0.5$, $n_4 = 0.25$, $n_5 = 2.0$, and $n_6 = 4.0$. The constants to be used in practice will depend on conditions and which model is favored. In particular, at a given flowrate high feed solids concentrations will substantially influence both cut-size (increase) and pressure drop (decrease). There is no general consensus, but in most applications the following values will give more realistic predictions: $n_1 = 1.54$, $n_2 = 0.43$, $n_3 = 0.72$, $n_4 = 0.22$, $n_5 = 2.0$, and $n_6 = 3.76$.

These relationships tell us that the diameter, flowrate, and pressure must be considered together. For example, cut-size cannot be scaled purely on cyclone diameter, as a new diameter will bring either a new flowrate or pressure or both. For example, if it is desired to scale to a larger cyclone at the same cut-size, then $d_{50c1} = d_{50c2}$ and:

$$D_{c2} = D_{c1}(P_2/P_1)^{n_4/n_3}$$

Classification efficiency can sometimes be improved by arranging several cyclones in series to retreat overflow, underflow, or both. Svarovsky and Thew (1984) have pointed out that if N cyclones with identical classification curves are arranged in series, each treating the overflow of the previous one, then the overall recovery of size d to the combined coarse product, $R_{d(T)}$, is given by:

$$R_{d(T)} = 1 - (1 - R_d)^N \qquad (3.28)$$

where R_d = recovery of size d in one cyclone.

(2) Sizing of Hydrocyclones—Arterburn Technique. Arterburn (1982) published a method, based on the performance of a "typical" Krebs cyclone, which allows for the cyclone size (i.e., diameter) to be estimated for a given application. Using a series of empirical and semi empirical relationships, the method relates the overflow P_{80} to the cut-point, d_{50c}, of a "base" cyclone, which is related to cyclone size. An example calculation illustrates the procedure (Example 3.4).

The cyclone size is the main choice for preliminary circuit design purposes. Variables, such as diameter of vortex finder, inlet, and apex, also affect separation (discussed in Section 3.4.5). Accordingly, most cyclones have replaceable vortex finders and apexes with different sizes available, and adjustments by operations will be made to provide the final design.

(3) Sizing of Hydrocyclones—MularJull Model. Mular and Jull (1980) developed empirical formulae from the graphical information for "typical" cyclones, relating d_{50c} to the operating variables for cyclones of varying diameter. A "typical" cyclone has an inlet area of about 7% of the cross-sectional area of the feed chamber, a vortex finder of diameter 35% ~ 40% of the cyclone diameter, and an apex diameter normally not less than 25% of the vortex-finder diameter.

The equation for the cyclone cut-point is:

$$d_{50c} = \frac{0.77 D_c^{1.875} \exp(-0.301 + 0.0945V - 0.00356V^2 + 0.0000684V^3)}{Q^{0.6}(S-1)^{0.5}} \qquad (3.29)$$

Equations such as these have been used in computer controlled grinding circuits to infer cut-points from measured data, but their use in this respect is declining with the increased use of online particle size monitors. Their value, however, remains in the design and optimization of circuits by the use of computer simulation, which greatly reduces the cost of assessing circuit options.

(4) Sizing of Hydrocyclones—Simulation Packages. A complement to the above methods of sizing cyclones is to use a simulation package such as those mentioned earlier (JKSimMet, MODSIM, or Limn). These packages incorporate empirical cyclone models such as those by Plitt and Nageswararao and can be used for optimizing processing circuits incorporating cyclones (Morrison and Morrell, 1998).

Example 3.4

Select (a) the size and (b) number of cyclones for a ball mill circuit for the following conditions:

(1) Target overflow particle size is 80% passing 90μm (i.e., P_{80} = 90μm).

(2) The pressure drop across the cyclone is 50kPa.

(3) Slurry feed rate to the cyclone is 1000m³/h.

(4) Slurry is 28% by volume of solids specific gravity 3.1(55% solids by weight).

Solution

(1) Size of cyclone (D_c).

Step 1: Estimate the cut-size for the application.

Arterburn gives a relationship between cut-size for the application and the P_{80}:

$$d_{50c}(\text{appl}) = 1.25 \times P_{80}$$

$$d_{50c}(\text{appl}) = 112.5 \mu m$$

Step 2: Determine a generic or base cut-size, $d_{50c}(\text{base})$ from the application, $d_{50c}(\text{appl})$:

$$d_{50c}(\text{base}) = \frac{d_{50}(\text{appl})}{C_1 \times C_2 \times C_3}$$

Where the C_3 are correction factors. The relationship changes the specific application to a cyclone operating under base conditions, which is therefore independent of the specific application operating parameters. (Olson and Turner (2002) introduce some additional correction factors reflective of cyclone design.)

The first correction term (C_1) is for the influence of the concentration of solids in the feed slurry:

$$C_1 = \left(\frac{53 - V}{53}\right)^{-1.43}$$

where V is the % solids by volume.

$$C_1 = \left(\frac{53 - 28}{53}\right)^{-1.43} \qquad C_1 = 2.93$$

The second correction (C_2) is for the influence of pressure drop (ΔP, kPa) across the cyclone, measured by taking the difference between the feed pressure and the overflow pressure:

$$C_2 = 3.27 \times (\Delta P)^{-0.28}$$

$$C_2 = 3.27 \times 50^{-0.28}$$

$$C_2 = 1.09$$

The final correction (C_3) is for the effect of solids specific gravity:

$$C_3 = \left(\frac{1.65}{\rho_s - \rho_l}\right)^{0.5}$$

where ρ_s and ρ_l are the specific gravity of solid and liquid (usually taken as water, $\rho_l = 1$). (Arterburn uses Stokes' law for particles of different densities but equal terminal velocities (Eq. (3.12)), with the reference being a quartz particle of specific gravity 2.65 in water.)

$$C_3 = \left(\frac{1.65}{3.1 - 1}\right)^{0.5} \qquad C_3 = 0.89$$

Thus, the $d_{50c}(\text{base})$ is:

$$d_{50c}(\text{base}) = \frac{112.5}{2.93 \times 1.09 \times 0.89}$$

$$d_{50c}(\text{base}) = 40 \text{cm}$$

Step 3: Estimate size (diameter) of cyclone from $d_{50c}(\text{base})$ using:

$$D_c = \left(\frac{d_{50c}(\text{base})}{2.84}\right)^{1.51}$$

$$D_c = 54\text{cm}$$

The cyclone sizes in Fig. 3.13 are in inches; from those shown the closest, but larger, selection is 26in (D-26). From the pressure drop-capacity chart (Fig. 3.13) a D-26 unit is suited to the target separation as the $d_{50c}(\text{appl})$ range is 70~150μm.

(2) Number of cyclones.

From Fig. 3.13 a D-26 cyclone can treat about 300m³/h. Thus to handle the target volumetric flowrate of 1000m³/h we need (1000/300), that is, four cyclones.

3.5 GRAVITATIONAL CLASSIFIERS

Gravitational classifiers are best suited for coarser classification and are often used as dewatering and washing equipment. They are simple to operate and have low energy requirements, but capital outlay is relatively high compared to cyclones. Gravitational classifiers can be further categorized into two broad groups, depending on the direction of flow of the carrying current: if the fluid movement is horizontal and forms an angle with the particle trajectory, the classification is called sedimentation classification; if the fluid movement and particle settling directions are opposite, the classification is called hydraulic or counter flow. Sedimentation, or horizontal current, classifiers are essentially of the free-settling type and accentuate the sizing function. On the other hand, hydraulic, or vertical current, classifiers are usually hindered-settling types and so increase the effect of density on the separation.

3.5.1 Sedimentation Classifiers

3.5.1.1 Nonmechanical Sedimentation Classifiers

As the simplest form of classifier, there is little attempt to do more than separate the solids from the liquid, and as such they are sometimes used as dewatering units in small-scale operations. Therefore, they are not suitable for fine classification or if a high separation efficiency is required. They are often used in the aggregate industry to deslime coarse sand products. The principle of the settling cone is shown in Fig. 3.17. The pulp is fed into the tank as a distributed stream, with the spigot discharge valve, S, initially closed. When the tank is full, overflow of water and slimes commences, and a bed of settled sand builds until it reaches the level shown. The spigot valve is now opened and sand is discharged at

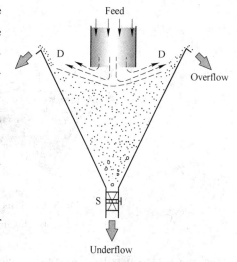

Fig. 3.17 Nonmechanical sedimentation classifier: settling cone operation

a rate equal to that of the input. Classification by horizontal current action takes place radially across zone D from the feed pipe to the overflow lip. The main difficulty in operation of such a device is the balancing of the sand discharge and deposition rates; it is virtually impossible to maintain a regular discharge of sand through an open pipe under the influence of gravity. Many different designs of cone have been introduced to overcome this problem (Taggart, 1945).

In the "Floatex" separator, which is essentially a hindered-settling classifier over a dewatering cone, automatic control of the coarse lower discharge is governed by the specific gravity of the teeter column. The use of the machine as a desliming unit and in upgrading coal and mica, as well as its possible application in closedcircuit classification of metalliferous ores, is discussed by Littler (1986).

3.5.1.2 Mechanical Sedimentation Classifiers

This term describes classifiers in which the material of lower settling velocity is carried away in a liquid overflow, and the material of higher settling velocity is deposited on the bottom of the unit and is transported upward against the flow of liquid by some mechanical means. The principle components of a mechanical classifier are shown in Fig. 3.18.

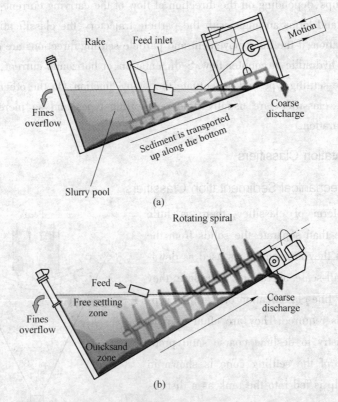

Fig. 3.18 Principle of mechanical classifier
(a) rake classifier, and (b) spiral classifier

Mechanical classifiers have seen use in closed-circuit grinding operations and in the classification of products from ore-washing plants (Chapter 2). In washing plants, they act more or less as sizing devices, as the particles are essentially unliberated and so are of similar density. In closed-circuit grinding, they have a tendency to return small dense particles to the mill, that is, the same as noted for cyclones. They have also been used to densify dense media (Chapter 5).

The pulp feed is introduced into the inclined trough and forms a settling pool in which coarse particles of high falling velocity quickly reach the bottom of the trough. Above this, coarse sand is a quicksand zone where hindered settling takes place. The depth and shape of this zone depends on the classifier action and on the feed pulp density. Above the quicksand is a zone of essentially free-settling material, comprising a stream of pulp flowing horizontally across the top of the quicksand zone from the feed inlet to the over flow weir, where the fines are removed.

The settled sands are conveyed up the inclined trough by a mechanical rake or by a helical screw. The conveying mechanism also serves to keep fine particles in suspension in the pool by gentle agitation and when the sands leave the pool they are slowly turned over by the raking action, thus releasing entrained slimes and water, increasing the efficiency of the separation. Washing sprays are often directed on the emergent sands to wash the released slimes back into the pool.

The rake classifier (Fig. 3.18(a)) uses rakes actuated by an eccentric motion, which causes them to dip into the settled material and to move it up the incline for a short distance. The rakes are then withdrawn, and return to the starting-point, where the cycle is repeated. The settled material is thus slowly moved up the incline to the discharge. In the duplex type, one set of rakes is moving up, while the other set returns; simplex and quadruplex machines are also made, in which there are one or four raking assemblies, respectively.

In spiral classifiers (Fig. 3.18(b), Fig. 3.19), a continuously revolving spiral moves the sands up the slope. They can be operated at steeper slopes than the rake classifier, in which the sands tend to slip back when the rakes are removed. Steeper slopes aid the drainage of sands, giving a cleaner, drier product. Agitation in the pool is less than in the rake classifier, which is important in separations of very fine material.

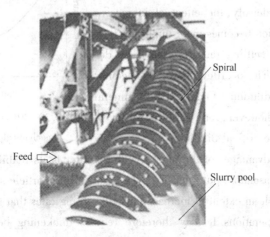

Fig. 3.19 Spiral classifier

The size at which the separation is made and the quality of the separation depend on a number of factors. Increasing the feed rate increases the horizontal carrying velocity and thus increases the size of the particles leaving in the overflow. The feed should not be introduced directly into the pool, as this causes agitation and releases coarse material from the hindered-settling zone, which may report to the overflow. The feed stream should be slowed down by spreading it on an apron, partially submerged in the pool, and sloped toward the sand discharge end, so that most of the kinetic energy is absorbed in the part of the pool furthest from the overflow.

The speed of the rakes or spiral determines the degree of agitation of the pulp and the tonnage rate of sand removal. For coarse separations, a high degree of agitation may be necessary to keep the coarse particles in suspension in the pool, whereas for finer separations, less agitation, and thus lower raking speeds, are required. It is essential, however, that the speed is high enough to transport the sands up the slope.

The height of the overflow weir is an operating variable in some mechanical classifiers. Increasing the weir height increases the pool volume, and hence allows more settling time and decreases the surface agitation, thus reducing the pulp density at overflow level, where the final separation is made. High weirs are thus used for fine separations.

Dilution of the pulp is the most important variable in the operation of mechanical classifiers. In closed-circuit grinding operations, ball mills rarely discharge at less than 65% solids by weight, whereas mechanical classifiers never operate at more than about 50% solids. Water to control dilution is added in the feed launder or onto the sand near the "V" of the pool. Water addition determines the settling rate of the particles. Increased dilution reduces the density of the weir overflow product, and increases free settling, allowing finer particles to settle out of the influence of the horizontal current. Therefore finer separations are produced, provided that the overflow pulp density is above a value known as the critical dilution, which is normally about 10% solids. Below this density, the effect of increasing rising velocity with dilution becomes more important than the increase in particle settling rates produced by decrease of pulp density. The overflow therefore becomes coarser with increasing dilution (Fig. 3.20). In mineral processing applications, however, very rarely is the overflow density less than the critical dilution.

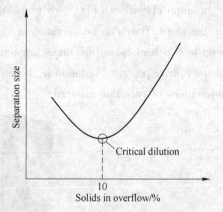

Fig. 3.20 Effect of dilution of overflow on mechanical classifier separation

One of the major disadvantages of the mechanical classifier is its inability to produce overflows of fine particle size at reasonable pulp densities. To produce fine particle separations, the pulp may have to be diluted to such an extent to increase particle settling rates that the overflow becomes too dilute for subsequent operations. It may therefore require thickening before mineral separation (concentration) can take place efficiently. This is undesirable as, apart from the capital cost and floor space of the thickener, oxidation of sulfide particles may occur in the thickener, which may affect subsequent processes, especially froth flotation.

3.5.2 Hydraulic Classifiers

Hydraulic classifiers are characterized by the use of water additional to that of the feed pulp, introduced so that its direction of flow opposes that of the settling particles. They normally consist of a series of sorting columns through which, in each column, a vertical current of water is rising and particles are settling out (Fig. 3.21 (a)).

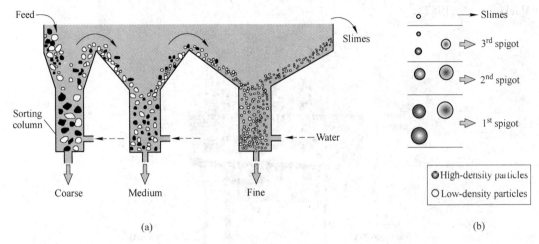

Fig. 3.21 (a) Principle of hydraulic classifier, and (b) spigot products

The rising currents are graded from a relatively high velocity in the first sorting column, to a relatively low velocity in the last, so that a series of underflow (spigot) products can be obtained, with the coarser, denser particles in the first underflow and progressively finer products in the subsequent underflows (Fig. 3.21 (b)). The finest fraction (slimes) overflows the final sorting column. The size of each successive vessel is increased, partly because the amount of liquid to be handled includes all the water used for classifying in the previous vessels and partly because it is desired to reduce, in stages, the surface velocity of the fluid flowing from one vessel to the next.

Hydraulic classifiers may be free-settling or hinderedsettling types. The free-settling types, however, are rarely used; they are simple and have high capacities, but are inefficient in sizing and sorting. They are characterized by the fact that each sorting column is of the same cross sectional area throughout its length.

The greatest use for hydraulic classifiers in the mineral industry is for sorting the feed to certain gravity concentration processes so that the size effect can be suppressed and the density effect enhanced (Chapter 4). Such classifiers are of the hindered-settling type. These differ from the free-settling classifiers in that the sorting column is constricted at the bottom in order to produce a teeter chamber. The hindered-settling classifier uses much less water than the free-settling type and is more selective in its action, due to the scouring action in the teeter chamber, and the buoyancy effect of the pulp, as a whole, on those particles which are to be rejected. Since the ratio of sizes of equally falling particles is high, the classifier is capable of performing a concentrating effect, and the first underflow product is normally richer in high-density material (often the valuable mineral) than the other products (Fig. 3.21). This is known as the added increment of the classifier and the first underflow product may in some cases be rich enough to be classed as a concentrate.

During classification the teeter bed tends to grow, as it is easier for particles to become entrapped in the bed rather than leave it. This tends to alter the character of the spigot discharge, as the density builds up. In multi-spigot hydrosizers, the teeter bed composition is automatically controlled. The Stokes hydrosizer (Fig. 3.22) was common in the Cornish (UK) tin industry (Mackie et al., 1987).

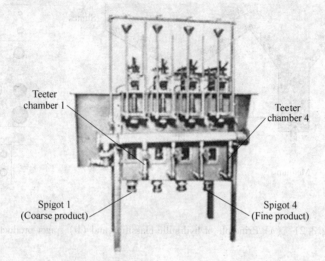

Fig. 3.22 Stokes multi-spigot hydrosizer

Each teeter chamber is provided at its bottom with a supply of water under constant head, which is used for maintaining a teetering condition in the solids that find their way down against the interstitial rising flow of water. Each teeter chamber is fitted with a discharge spigot that is, in turn, connected to a pressure-sensitive valve so that the classifying conditions set by the operator can be accurately controlled.

The valve may be hydraulically or electrically controlled; in operation it is adjusted to balance the pressure set up by the teetering material. The concentration of solids in a particular compartment can be held steady in spite of the normal variations in feed rate taking place from time to time. The rate of discharge from each spigot will, of course, change in sympathy with these variations, but since these changing tendencies are always being balanced by the valve, the discharge will take place at a nearly constant density. For a quartz sand this is usually about 65% solids by weight, but is higher for heavier minerals.

REFERENCES

Anon, 1984. Classifiers Part 2: some of the major manufacturers of classification equipment used in mineral processing. Mining Mag. 151 (Jul.), 40-44.

Arterburn, R. A., 1982. Sizing, and selection of hydrocyclones. In: Mular, A. L., Jergensen, G. V. (Eds.), Design and Installation of Comminution Circuits, vol. 1. AIME, New York, NY, USA, 597-607.

(Chapter 32).

Asomah, A. K., Napier-Munn, T. J., 1997. An empirical model of hydrocyclones incorporating angle of cyclone inclination. Miner. Eng. 10(3), 339-347.

Austin, L. G., et al., 1984. Process Engineering of Size Reduction: Ball Milling. SME, New York, NY, USA.

Bahar, A., et al., 2014. Lesson learned from using column flotation cells as roughers: the Miduk copper concentrator plant case. Proceedings of the 27th International Mineral Processing Congress (IMPC), Chapter 6, Paper C0610, Santiago, Chile, 1-12.

Bradley, D., 1965. The Hydrocyclone: International Series of Monographs in Chemical Engineering. first ed. Pergamon Press Ltd., Oxford, UK.

Brennan, M. S., et al., 2003. Towards a new understanding of the cyclone separator. Proceedings of the 22nd International Mineral Processing Congress (IMPC), vol. 1, Cape Town, South Africa, 378-385.

Bourgeois, F., Majumder, A. K., 2013. Is the fish-hook effect in hydrocyclones a real phenomenon? . Powder Technol. 237, 367-375.

Chang, Y.-F., et al., 2011. Particle flow in a hydrocyclone investigated by positron emission particle tracking. Chem. Eng. Sci. 66 (18), 4203-4211.

Cirulis, D., Russell, J., 2011. Cyclone monitoring system improves operations at KUC's Copperton concentrator. Eng. Min. J. 212 (10), 44-49.

Del Villar, R., Finch, J. A., 1992. Modelling the cyclone performance with a size dependent entertainment factor. Miner. Eng. 5 (6), 661-669.

Dundar, H., et al., 2014. Screens and cyclones in closed grinding circuits. Proceedings of the 27th International Mineral Processing Congress (IMPC), Chapter 16, Paper C1607, Santiago, Chile, 1-11.

Flintoff, B. C., et al., 1987. Cyclone modelling: a review of present technology. CIM Bull. 80 (905), 39-50.

Hand, P., Wiseman, D., 2010. Addressing the envelope. J. S. Afr. Inst. Min. Metall. 110, 365-370.

Heiskanen, K., 1993. Particle Classification. Chapman & Hall, London, UK.

Kawatra, S. K., Eisele, T. C., 2006. Causes and significance of inflections in hydrocyclone efficiency curves. In: Kawatra, S. K. (Ed.), Advances in Comminution. SME, Littleton, CO, USA, 131-147.

Kawatra, S. K., Seitz, R. A., 1985. Technical note: calculating the particle size distribution in a hydrocyclone overflow product for simulation purposes. Miner. Metall. Process. 2 (Aug.), 152-154.

Kelsall, D. F., 1953. A further study of the hydraulic cyclone. Chem. Eng. Sci. 2 (6), 254-272.

King, R. P., 2012. Modeling and Simulation of Mineral Processing Systems. Second ed. SME, Englewood, CO, USA.

Kraipech, W., et al., 2006. The performance of the empirical models on industrial hydrocyclone design. Int. J. Miner. Process. 80, 100-115.

Laplante, A. R., Finch, J. A., 1984. The origin of unusual cyclone performance curves. Int. J. Miner. Process. 13 (1), 1-11.

Leith, D., Licht, W., 1972. The collection efficiency of cyclone type particle collector: a new theoretical approach. AIChE Symp. Series (Air-1971). 68 (126), 196-206.

Littler, A., 1986. Automatic hindered-settling classifier for hydraulic sizing and mineral beneficiation. Trans. Inst. Min. Metall., Sec. C. 95, C133-C138.

Lynch, A. J., Rowland, C. A., 2005. The History of Grinding. SME, Littleton, CO, USA.

Lynch, A. J., et al., 1974. The influence of hydrocyclone diameter on reduced-efficiency curves. Int. J. Miner. Process. 1 (2), 173-181.

Lynch, A. J., et al., 1975. The influence of design and operating variables on the capacities of hydrocyclone classifiers. Int. J. Miner. Process. 2 (1) ,29-37.

Mackie, R. I. , et al. , 1987. Mathematical model of the Stokes hydrosizer. Trans. Inst. Min. Metall. , Sec. C. 96, C130-C136.

Morrison, R. D. , Freeman, N. , 1990. Grinding control development at ZC Mines. Proc. AusIMM. 295 (2) ,45-49.

Morrison, R. D. , Morrell, S. , 1998. Comparison of comminution circuit energy efficiency using simulation. Miner. Metall. Process. 15 (4) ,22-25.

Mular, A. L. , Jull, N. A. , 1980. The selection of cyclone classifiers, pumps and pump boxes for grinding circuits, Mineral Processing Plant Design. second ed. AIMME, New York, NY, USA, 376-403.

Nageswararao, K. , 1995. Technical note: a generalised model for hydrocyclone classifiers. Proc. AusIMM. 300 (2) ,21.

Nageswararao, K. , 2014. Comment on: 'Is the fish-hook effect in hydrocyclones a real phenomenon?' by F. Bourgeois and A. K Majumder [Powder Technology 237 (2013) 367-375]. Powder Technol. 262,194-197.

Nageswararao, K. , et al. , 2004. Two empirical hydrocyclone models revisited. Miner. Eng. 17 (5) ,671-687.

Napier-Munn, T. J. , et al. , 1996. Mineral Comminution Circuits: Their Operation and Optimisation. Chapter 12, Julius Kruttschnitt Mineral Research Centre (JKMRC), The University of Queensland, Brisbane, Australia.

Narasimha, M. , et al. , 2005. CFD modelling of hydrocyclone—prediction of cut-size. Int. J. Miner. Process. 75 (1-2) ,53-68.

Nowakowski, A. F. , et al. , 2004. Application of CFD to modelling of the flow in hydrocyclones. Is this a realizable option or still a research challenge? Miner. Eng. 17 (5) ,661-669.

Olson, T. J. , Turner, P. A. , 2002. Hydrocyclone selection for plant design. In: Mular, A. L. , et al. , (Eds.) , Mineral Processing Plant Design, Practice, and Control, vol. 1. Vancouver, BC, Canada, 880-893.

Plitt, L. R. , 1976. A mathematical model of the hydrocyclone classifier. CIM Bull. 69 (Dec.) ,114-123.

Rao, T. C. , et al. , 1976. Influence of feed inlet diameter on the hydrocyclone behaviour. Int. J. Miner. Process. 3 (4) ,357-363.

Renner, V. G. , Cohen, H. E. , 1978. Measurement and interpretation of size distribution of particles within a hydrocyclone. Trans. Inst. Min. Metall. , Sec. C. 87 (June) , C139-C145.

Sasseville, Y. , 2015. Study on the impact of using cyclones rather than screens in the grinding circuit at Niobec Mine. Proceedings of the 47th Annual Meeting of the Canadian Mineral Processors Conference, Ottawa, ON, Canada, 118-126.

Smith, V. C. , Swartz, C. L. E. , 1999. Development of a hydrocyclone product size soft-sensor. In: Hodouin, D. et al. , (Eds.) , Proceedings of the International Symposium on Control and Optimization in Minerals, Metals and Materials Processing. 38th Annual Conference of Metallurgists of CIM: Gateway to the 21st Century. Quebec City, QC, Canada, 59-70.

Stokes, G. G. , 1891. Mathematical and Physical Papers, vol. 3. Cambridge University Press.

Svarovsky, L. , Thew, M. T. , 1992. Hydrocyclones: Analysis and Applications. Springer Science 1 Business Media, LLC, Technical Communications (Publishing) Ltd. , Letchworth, England.

Swanson, V. F. , 1989. Free and hindered settling. Miner. Metall. Process. 6 (Nov.) ,190-196.

Taggart, A. F. , 1945. Handbook of Mineral Dressing, Ore and Industrial Minerals. John Wiley & Sons, Chapman & Hall, Ltd, London, UK.

Wang, Zh. -B. , et al. , 2008. Experimental investigation of the motion trajectory of solid particles inside the hydrocyclone by a Lagrange method. Chem. Eng. J. 138 (1-3) ,1-9.

Westendorf, M. , et al. , 2015. Managing cyclones: A valuable asset, the Copper Mountain case study. Min. Eng. 67 (6) ,26-41.

Chapter 4   Gravity Concentration

4.1 INTRODUCTION

Gravity concentration is the separation of minerals based upon the difference in density. Techniques of gravity concentration have been around for millennia. Some believe that the legend of the Golden Fleece from Homer's Odyssey was based upon a method of gold recovery, which was to place an animal hide (such as a sheep's fleece) in a stream containing alluvial gold; the dense gold particles would become trapped inn the fleece and be recovered (Agricola,1556). In the various gold rushes of the nineteenth century, many prospectors used gold panning as a means to make their fortune-another ancient method of gravity concentration. Gravity concentration or density-based separation methods,declined in impor-tance in the first half of the twentieth century due to the development of froth flotation, which allowed for the selective treatment of low-grade complex ores. They remain,however,the main concentrating methods for iron and tungsten ores and are used exten-sively for treating tin ores,coal,gold,beach sands,and many industrial minerals.

In recent years,mining companies have reevaluated gravity systems due to increasing costs of flotation reagents, the relative simplicity of gravity processes, and the fact that they produce comparatively little environ mental impact. Modern gravity techniques have proved efficient for concentration of minerals having particl esizes down to the 50μm range and, when coupled with improved pumping technology and instrumentation, have been incorporated in high-capacity plants (Holland-Batt,1998). In many cases a high proportion of the mineral inan ore body can be preconcentrated effectively by gravity separation systems; the amount of reagents and energyused can be cut significantly when the more ex pensive methods are restricted to the processing of a gravity concentrate. Gravity separation at coarse sizes (as soon asliberation is achieved) can also have significant advantages for later treatment stages, due to decreased surface area, more efficient dewatering, and the absence of adsorbed chemicals, which could interfere with further processing.

4.2 PRINCIPLES OF GRAVITY CONCENTRATION

Gravity concentration methods separate minerals of different specific gravity by their relative movement inresponse to gravity and one or more other forces, the latter often being the resistance to motion offered by a viscous fluid, such as water or air.

It is essential for effective separation that a marked density difference exists between the mineral and the gangue. Some idea of the type of separation possible can be gained from the concentration criterion, $\Delta \rho$:

$$\Delta \rho = \frac{\rho_h - \rho_f}{\rho_l - \rho_f} \qquad (4.1)$$

where ρ_h is the density of the heavy mineral, ρ_l is the density of the light mineral, and ρ_f is the density of the fluid medium.

In general terms, if the quotient has a magnitude greater than 2.5, then gravity separation is relatively easy, with the efficiency of separation decreasing as the value of the quotient decreases. To give an example, if gold is being separated from quartz using water as the carrier fluid, the concentration criterion is 11.1 (density of gold being $19300 kg/m^3$, quartz being $2650 kg/m^3$), which is why panning for gold has been so successful.

The motion of a particle in a fluid is dependent notonly on its specific gravity, but also on its size (Chapter 3)—large particles will be affected more than smaller ones. The efficiency of gravity processes therefore increases with particle size, and the particles should be sufficiently coarse to move in accordance with Newton's law (Eq. (3.8)). Particles small enough that their movement is dominated mainly by surface friction respond relatively poorly to commercial high-capacity gravity methods. In practice, close size control of feeds to gravity processes is required to reduce the size effect and make the relative motion of the particles specific gravity-dependent. The incorporation of enhanced gravity concentrators, which impart additional centrifugal acceleration to the particles (Section 4.4), have been utilized in order to overcome some of the drawbacks of fine particle processing. Table 4.1 gives the relative ease of separing minerals using gravity techniques, based upon the pticle size and concentration criterion (Anon., 2011).

Table 4.1 Dependence on Concentration Criterion (Eq. (4.1)) for Separations

Concentration criterion	Separation	Useful for
2.5	Relatively easy	To 75μm
1.75~2.5	Possible	To 150μm
1.5~1.75	Difficult	To 1.7mm
1.25~1.5	Very difficult	
<1.25	Not possible	

4.3 GRAVITATIONAL CONCENTRATORS

Many machines have been designed and built to effect separation of minerals by gravity (Burt, 1985). The densemedium separation (DMS) process is used to preconcentrate crushed material prior to grinding and will be considered separately in the next chapter (Chapter 5). Design and optimization of gravity circuits is discussed by Wells (1991) and innovations in gravity separation are reviewed by Honaker et al. (2014).

It is essential for the efficient operation of all gravity separators that the feed is carefully prepared. Grinding is particularly important to provide particles of adequate liberation; successive regrinding of middlings is required inmost operations. Primary grinding should be performed where possible in open-circuit rod mills, but if fine grinding is required, closed-circuit ball milling should be used, preferably with screens closing the circuits rather than hydrocyclones, in order to reduce selective overgrinding of heavy friable valuable minerals. Other methods of comminution, such as semi-autogeneous grindingmills and high-pressure grinding rolls, may find application in preparing gravity feeds.

Gravity separators are sensitive to the presence of slimes (ultrafine particles), which increase the viscosity of the slurry and hence reduce the sharpness of separation, and obscure visual cut-points for operators. It has been common practice to remove particles less than about $10\mu m$ from the feed (i. e. , deslime), and divert this fraction to the tailings, which can incur considerable loss of values. Desliming is often achieved by the use of hydrocyclones, although hydraulic classifiers may be preferablein some cases since the high shear forces produced in hydrocyclones tend to cause degradation of friable minerals (and create more loss of slimed values).

The feed to jigs and spirals should, if possible, be screened before separation takes place, each fraction being treated separately. In most cases, however, removal of the oversize by screening, in conjunction with desliming, is adequate. Processes which employ flowing-film separation, such as shaking tables, should always bepreceded by good hydraulic classification in multi-spigothydrosizers (Chapter 3).

Although most slurry transportation is achieved by centrifugal pumps and pipelines, as much as possible use should be made of natural gravity flow; many old gravity concentrators were built on hillsides to achieve this. Reduction of slurry pumping to a minimum not only lowers energy consumption, but also reduces slimes production in the circuit. To minimize degradation of friable minerals, pumping velocities should be as low as possible, consistent with maintaining the solids in suspension.

One of the most important aspects of gravity circuit operations is correct water balance within the plant. Almost all gravity concentrators have an optimum feedpulp density, and relatively little deviation from this density causes a rapid decline in efficiency. Accurate pulpdensity control is therefore essential, and this is most important on the raw feed. Automatic density control should be used where possible, and the best way of achieving this is by the use of nucleonic density gauges controlling the water addition to the newfeed. Although such instrumentation is expensive, it is usually economic in the long term. Control of pulp density within the circuit can be made by the use of settling cones (Chapter 3) preceding the gravity device. These thicken the pulp, but the overflow often contains solids, and should be directed to a central large sump or thickener. For substantial increase in pulp density, hydrocyclones or thickeners may be used. The latter are the more expensive, but produce less particle degradation and also provide substantial surge capacity. It is usually necessary to recycle water in most plants, so adequate thickener or cyclone capacity should be provided, and slimes build-up in the recycled water must be minimized.

If the ore contains an appreciable amount of sulfideminerals then, because they are relatively dense and tend to report with the heavy product, they need to be removed. If the primary grind is finer than ca. 300 pm, the sulfides should be removed by froth flotation prior to gravity concentration. If the primary grind is too coarse for effective sulfide flotation, then the gravity concentrate must be reground prior to removal of the sulfides. The sulfide flotation tailing is then usually cleaned by further gravity concentration.

The final gravity concentrate often needs cleaning by magnetic separation, leaching, or some other method, in order to remove certain mineral contaminants. For instance, at the South Crofty tin mine in Cornwall, the gravity concentrate was subjected to cleaning by magnetic separators, which removed wolframite from the cassiterite product.

4.3.1 Jigs

Jigging is one of the oldest methods of gravity concentration, yet the basic principles have only recently been understood. A mathematical model developed by Jonkerset al. (2002) predicts jig performance on a size-by-density basis; Mishra and Mehrotra (1998) developed discrete element method models of particle motion in a jig; and Mishra and Mehrotra (2001) and Xia et al. (2007) have developed a computational fluid dynamics model of coalstratification in a jig.

In the jig, the separation of minerals of different specific gravity is accomplished in a particle bed, which is fluidized by a pulsating current of water, producing stratification based upon density. The bed may be a specific mineral added to and retained in the jig, called ragging, composed of a certain density and shape through which the dense particles penetrate and the light particles passover the top. The aim is to dilate the bed and to control the dilation so that the heavier, smaller particles penetrate the interstices of the bed and the larger high-specific gravity particles fall under a condition similar to hindered settling (Lyman, 1992).

On the pulsion stroke the bed is normally lifted as a mass, then as the velocity decreases it tends to dilate, the bottom particles falling first until the whole bed is loosened. On the suction stroke it then closes slowly again, and this is repeated at every stroke. Fine particles tend to pass through the interstices after the large ones have become immobile. The motion can be obtained either by using a fixed sieve jig, and pulsating the water, or by employing a moving sieve, as in the simple hand-jig shown in Fig. 4.1.

The jig is normally used to concentrate relatively coarse material and, if the feed is fairly close-sized (e.g., 3~10mm), it is not difficult to achieve good separation of a fairly narrow specific gravity range in minerals in the feed (e.g., fluorite, s.g. 3.2, from quartz, s.g. 2.7). When the specific gravity difference is large, good concentration is possible with a wider size range. Many large jig circuits are still operated in coal, cassiterite, tungsten, gold, barytes, and iron ore concentrators. They have a relatively high unit capacity on classified feed and can achieve good recovery of values down to 150pm and acceptable recoveries often down to 75pm. High proportions of fine sand and slime interfere with performance and the fines content should be controlled to provide optimum bed conditions.

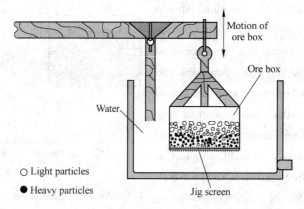

Fig. 4.1 Jigging action: by moving box up and down the particles segregate by density
(Adapted from Priester et al. (1993))

4.3.1.1 Jigging Action

It was shown in Chapter 3 that the equation of motion of a particle settling in a viscous fluid is:

$$m \frac{dx}{dt} = mg - m'g - D \tag{4.2}$$

where m is the mass of the mineral particle, dx/dt is the acceleration, g is the acceleration due to gravity, m' is the mass of displaced fluid, and D is the fluid resistance due to the particle movement.

At the beginning of the particle movement, since the velocity x is very small, D can be ignored, as it is a function of velocity:

Therefore

$$\frac{dx}{dt} = \left(\frac{m - m'}{m}\right) g \tag{4.3}$$

and since the particle and the displaced fluid are of equal volume:

$$\frac{dx}{dt} = \left(\frac{\rho_s - \rho_f}{\rho_s}\right) g \tag{4.4}$$

where ρ_s and ρ_f are the respective densities of the solid and the fluid.

The initial acceleration of the mineral grains is thus independent of size and dependent only on the densities of the solid and the fluid. Theoretically, if the duration of fall is short enough and the repetition of fall frequent enough, the total distance travelled by the particles will be affected more by the differential initial acceleration, and therefore by density, than by their terminal velocities, and therefore by size. In other words, to separate small heavy mineral particles from large light particles, a short jigging cycle is necessary. Although relatively short, fast strokes are used to separate fine minerals, more control and better stratification can be achieved by using longer, slower strokes, especially with the coarser particle sizes. It is therefore good practice to screen the feed to jigs into different size ranges and treat these separately. The effect of differential initial acceleration is shown in Fig. 4.2.

If the mineral particles are examined after a long time, they will have attained their terminal velocities and will be moving at a rate dependent on their specific gravity and size. Since the bed is really a loosely packed mass with interstitial water providing a very thick suspension of high density, hindered-settling conditions prevail, and the settling ratio of heavy to light minerals is higher than that for free settling (Chapter 3). Fig. 4.3 shows the effect of hindered settling on the separation.

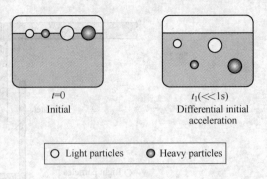

Fig. 4.2 Differential initial acceleration

The upward flow can be adjusted so that it overcomes the downward velocity of the fine light particles and carries them away, thus achieving separation. It can be increased further so that only large heavy particles settle, but it is apparent that it will not be possible to separate the small heavy and large light particles of similar terminal velocity.

Hindered settling has a marked effect on the separation of coarse minerals, for which longer, slower strokes should be used, although in practice, with coarser feeds, it is unlikely that the larger particles have sufficient time to reach their terminal velocities.

At the end of a pulsion stroke, as the bed begins to compact, the larger particles interlock, with the smaller particles still moving downward through the interstices under the influence of gravity. The fine particles may not settle as rapidly during this consolidation trickling phase (Fig. 4.4) as during the initial acceleration or suspension, but if consolidation trickling can be made to last long enough, the effect, especially in the recovery of the fine heavy minerals, can be considerable.

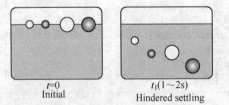

Fig. 4.3 Hindered settling

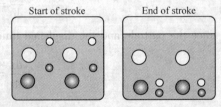

Fig. 4.4 Consolidation trickling

Fig. 4.5 shows an idealized jigging process by the described phenomena.

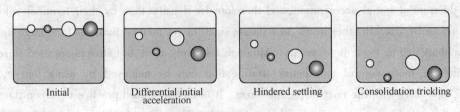

Fig. 4.5 Ideal jigging process

In the jig the pulsating water currents are caused by a piston having a movement that is a harmonic waveform (Fig. 4. 6). The vertical speed of flow through the bed is proportional to the speed of the piston. When this speed is greatest, the speed of flow through the bed is also greatest (Fig. 4. 7).

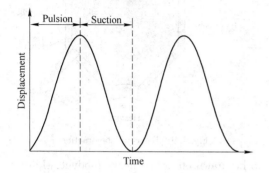 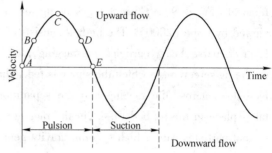

Fig. 4. 6 Movement of the piston in a jig   Fig. 4. 7 Speed of flow through bed during jig cycle

The upward speed of flow increases after point A, the beginning of the cycle. As the speed increases, the particles will be loosened and the bed will be forced open, or dilated. At, say, point B, the particles are in the phase of hindered settling in an upward flow, and since the speed of flow from B to C still increases, the fine particles are pushed upward by the flow. The chance of them being carried along with the top flow into the low-density product (often the tailings) is then at its greatest. In the vicinity of D, first the coarser particles and later the remaining fine particles will fall back. Due to the combination of initial acceleration and hindered settling, it is mainly the coarser particles that will lie at the bottom of the bed.

At E, the point of transition between the pulsion and the suction stroke, the bed will be compacted. Consolidation trickling can now occur to a limited extent. In a closely sized ore, the heavy particles can now penetrate only with difficulty through the bed and may be lost to the low-density stream. Severe compaction of the bed can be reduced by the addition of hutch water, a constant volume of water, which creates a constant upward flow through the bed. This flow, coupled with the varying flow caused by the piston, is shown in Fig. 4. 8. Thus suction is reduced by hutch-water addition and is reduced in duration; by adding a large quantity of water, the suction

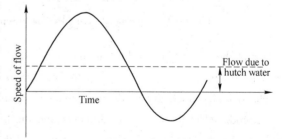

Fig. 4. 8 Effect of hutch water on flow through bed

may be entirely eliminated. The coarse ore then penetrates the bed more easily and the horizontal transport of the feed over the jig is also improved. However, fines losses will increase, partly because of the longer duration of the pulsion stroke, and partly because the added water increases the speed of the top flow.

4.3.1.2 Types of Jig

Essentially, the jig is an open tank filled with water, with a horizontal jig screen at the top supporting the jig bed, and provided with a spigot in the bottom, or hutch compartment, for "heavies" removal (Fig. 4.9). Current types of jig are reviewed by Cope (2000). The jig bed consists of a layer of coarse, heavy particles, the ragging, placed on the jig screen on to which the slurry is fed. The feed flows across the ragging and the separation takes place in the jig bed (i.e., in the ragging), so that particles with a high specific gravity penetrate through the ragging and then pass the screen to be drawn off as the heavy product, while the light particles are carried away by the cross-flow to form the light product. The type of ragging material, particle density, size, and shape, are important factors. The harmonic motion produced by the eccentric drive is supplemented by a large amount of continuously supplied hutch water, which enhances the upward and diminishes the downward velocity of the water (Fig. 4.8). The combination of actions produces the segregation depicted in Fig. 4.10.

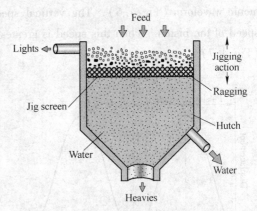

Fig. 4.9 Basic jig construction

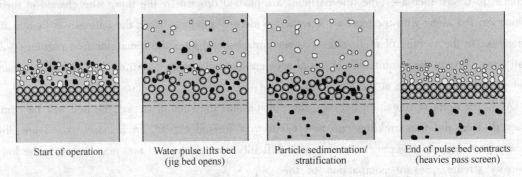

Fig. 4.10 Sequence of events leading to separation of heavies from lights in a jig

One of the oldest types is the Harz jig (Fig. 4.11) in which the plunger moves up and down vertically in a separate compartment. Up to four successive compartments are placed in series in the hutch. A high-gradeheavy product is produced in the first compartment, successively lower grades being produced in the other compartments, and the light product overflowing the final compartment. If the feed particles are larger than the apertures of the screen, jigging "over the screen" is used, an- the concentrate grade is partly governed by the thicknes of the bottom layer, determined by the rate of withdrawal through the concentrate discharge port.

The Denver mineral jig (Fig. 4.12) is widely used, especially for removing heavy minerals from closed grinding circuits, thus preventing over-grinding (Chapter 3). The rotary water valve can be adjusted so as toopen at any desired part of the jig cycle, synchronization between the valve and the

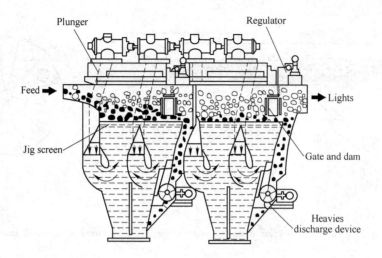

Fig. 4.11 Harz jig (Adapted from The Great Soviet Encyclopedia, 3rd Edn.)

plungers being achieved by a rubber timing belt. By suitable adjustment of the valve, any desired variation can be achieved, from complete neutralization of the suction stroke with hydraulic water to a full balance between suction and pulsion.

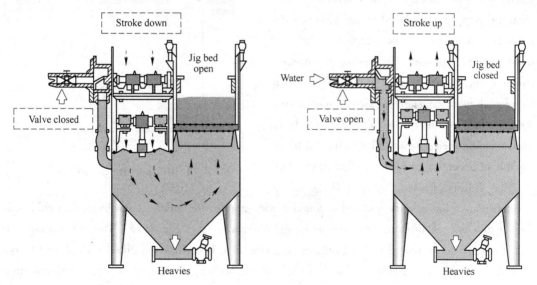

Fig. 4.12 Denver mineral jig

Conventional mineral jigs consist of square or rectangular tanks, sometimes combined to form two, three, or four cells in series. In order to compensate for the increase in cross-flow velocity over the jig bed, caused by the addition of hutch water, trapezoidal-shaped jigs were developed. By arranging these as sectors of a circle, the modular circular, or radial, jig was introduced, in which the feed enters in the center and flows radially over the jig bed (and thus the cross-flow velocity is decreasing) toward the light product discharge at the circumference (Fig. 4.13).

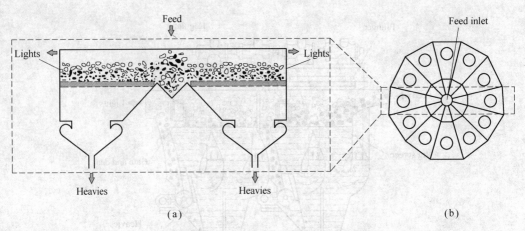

Fig. 4.13 (a) Cross section through a circular jig, and (b) radial jig up to 12 modules

The advantage of the circular jig is its large capacity. Since their development in 1970, IHC Radial Jigs were installed on most newly built tin dredges in Malaysia and Thailand. In the IHC jig, the harmonic motion of the conventional eccentric-driven jig is replaced by an asymmetrical sawtooth movement of the diaphragm, with a rapid upward, followed by a slow downward, stroke (Fig. 4.14). This produces a much larger and more constant suction stroke, giving the finer particles more time to settle in the bed, thus reducing their loss to tailings, the jig being capable of accepting particles as fine as 60μm.

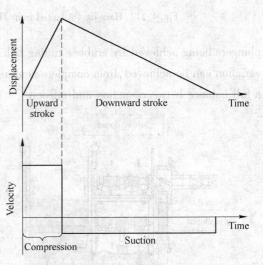

Fig. 4.14 IHC jig drive characteristics

The InLine Pressure Jig (IPJ) is a development in jig technology that has found a wide application for the recovery of free gold, sulfides, native copper, tin/tantalum, diamonds, and other minerals (Fig. 4.15). The IPJ is unique in that it is fully encapsulated and pressurized, allowing it to be completely filled with slurry (Gray, 1997). It combines a circular bed with a vertically pulsed screen. Length of stroke and pulsation frequency, as well as screen aperture, can all be altered to suit the application. IPJs are typically installed in grinding circuits, where their low water requirements allow operators to treat the full circulating load, maximizing recovery of liberated values. Both heavy and light products are discharged under pressure.

Jigs are widely used in coal cleaning (also referred to as "coal washing") and are preferred to the more expensive DMS when the coal has relatively little middlings, or "near-gravity" material. No feed preparation is required, as is necessary with DMS, and for coals that are easily washed, that is, those consisting predominantly of liberated coal and denser "rock" particles, the lack of close density control is not a disadvantage.

Two types of air-pulsated jig-Baum and Batac—are used in the coal industry. The standard Baum jig (Fig. 4.16), with some design modifications (Green, 1984), has been used for nearly 100 years and is still the dominant device.

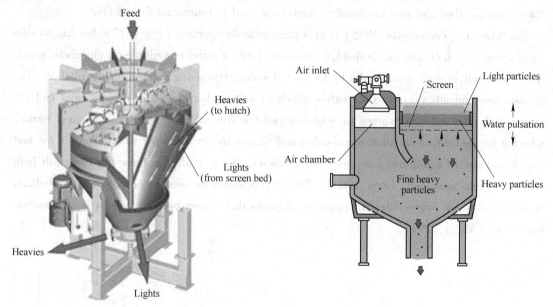

Fig. 4.15 The Gekko Systems IPG
(Courtesy Gekko Systems)

Fig. 4.16 Baum jig

Air under pressure is forced into a large air chamberon one side of the jig vessel, causing pulsation and suction to the jig water, which in turn causes pulsation and suction through the screen plates upon which the raw coalis fed, thus causing stratification. Various methods are used to continuously separate the refuse (heavy non-coalmatter) from the lighter coal product, and all the modern Baum jigs are fitted with some form of automatic refuse extraction. One form of control incorporates a float immersed in the bed of material. The float is suitably weighted to settle on the dense layer of refuse moving across the screen plates. An increase in the depth of refuse raises the float, which automatically controls the refuse discharge, either by adjusting the height of a moving gate, or by controlling the pulsating water, which lifts the rejects over a fixed weir plate (Wallace, 1979). This system is reported to respond quickly and accurately.

It is now commonplace for an automatic control system to determine the variations in refuse bed thickness by measuring the differences in water pressure under the screen plates arising from the resistance offered to pulsation. The JigScan control system (developed at the Julius Kruttschnitt Mineral Research Centre) measures bed conditions and pulse velocity many times within the pulseusing pressure sensing and nucleonic technology (Loveday and Jonkers, 2002). Evidence of a change in the pulse is an indicator of a problem, allowing the operator to take corrective action. Increased yields of greater than 2% have been reported for JigScan-controlled jigs.

In many situations, the Baum jig still performs satisfactorily, with its ability to handle large tonnages (up to 1000t/h) of coal of a wide size range. However, the distribution of the stratification force, being on one side of the jig, tends to cause unequal force along the width of jig screen and therefore uneven stratification and some loss in the efficiency of separation of the coal from its heavier impurities. This tendency is not so important in relatively narrow jigs, and in the United States multiple float and gate mechanisms have been used to counteract the effects.

The Batac jig (Zimmerman, 1975) is also pneumatically operated (Fig. 4.17), but has no side air chamber like the Baum jig. Instead, it is designed with a series of multiple air chambers, usually two to a cell, extending under the jig for its full width, thus giving uniform air distribution. The jig uses electronically controlled air valves which provide a sharp cutoff of the air input and exhaust. Both inlet and outlet valves are infinitely variable with regard to speed and length of stroke, allowing for the desired variation in pulsation and suction by which proper stratification of the bed may be achieved for differing raw coal characteristics. As a result, the Batac jig can wash both coarse and fine sizes well (Chen, 1980). The jig has also been used to produce high-grade lump ore and sinter-feed concentrates from iron ore deposits that cannot be upgraded by heavy-medium techniques (Miller, 1991).

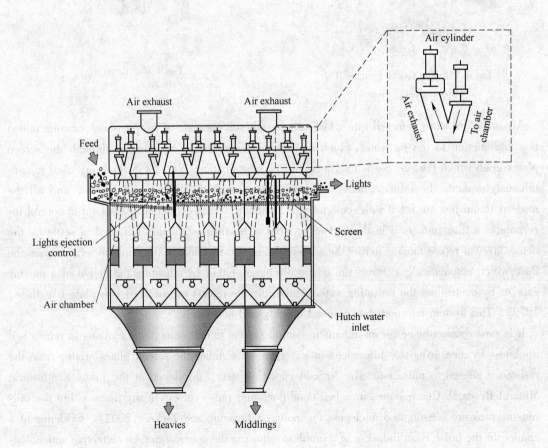

Fig. 4.17 Batac jig

4.3.2 Spirals

Spiral concentrators have found many varied applications in mineral processing, but perhaps their most extensive application has been in the treatment of heavy mineral sand deposits, such as those carrying ilmenite, rutile, zircon, and monazite, and in recent years in the recovery of fine coal.

The Humphreys spiral was introduced in 1943, its first commercial use being on chrome-bearing sands. It is composed of a helical conduit of modified semicircular cross section. Feed pulp of between 15% and 45% solids by weight and in the size range from 3mm to 75μm is introduced at the top of the spiral. As it flows spirally downward, the particles stratify due to the combined effect of centrifugal force, the differential settling rates of the particles, and the effect of interstitial trickling through the flowing particle bed. The result of this action is depicted in Fig. 4.18. Fig. 4.19 shows the stratification across a spiral trough, with the darker heavy mineral toward the center, with the band becoming increasingly lighter radially where the less dense material flows.

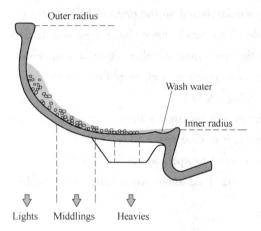

Fig. 4.18 Cross section of spiral stream

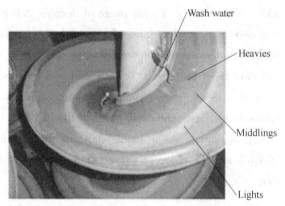

Fig. 4.19 Example of stratification across a spiral trough, with dense, dark material in the center and lighter, less dense extending out radially (Courtesy Multotec)

These mechanisms are complex, being much influenced by the slurry density and particle size. Mills (1980) reported that the main separation effect is due to hindered settling, with the largest, densest particles reporting preferentially to the band that forms along the inner edge of the stream. Bonsu (1983), however, reported that the net effect is reverse classification, the smaller, denser particles preferentially entering this band.

Determination of size-by-size recovery curves of spiral concentrators has shown that both fine and coarse dense particles are lost to the light product, the loss of coarse particles being attributed to the Bagnold force (Bazinet al. ,2014). Some of the complexities of the spiral concentrator operation arise from the fact that there is not one flow pattern, but rather two: a primary flow down the spiral and a secondary flow across the trough flowing outward at the top of the stream and inward at the bottom(Fig. 4.20) (Holland-Batt and Holtham,1991).

Ports for the removal of the higher specific gravity particles are located at the lowest points in the cross section. Wash water is added at the inner edge of the stream and flows outwardly across the concentrate band to aid in flushing out entrapped light particles. Adjustable splitters control the width of the heavy product band removed at the ports. The heavy product taken via descending ports is of a progressively decreasing grade, with the light product discharged from the lower end of the spiral. Splitters at the end of the spiral are often used to give three products: heavies, lights, and middlings, giving possibilities for recycle and retreatment in other spirals (e.g., roughercleaner spiral combination) or to feed other separation units. Incorporating automatic control of splitter position is being developed (Zhang et al., 2012).

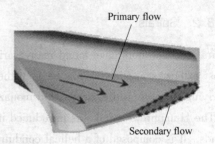

Fig. 4.20 Schematic diagram of the primary and secondary flows in a spiral concentrator

Until the last 20 years or so, all spirals were quite similar, based on the original Humphreys design. Today there is a wide range of designs available. Two developments have been spirals with only one heavy product take-off port at the bottom of the spiral, and the elimination of wash water. Wash waterless-spirals reportedly offer lower cost, easier operation, and simplified maintenance, and have been installed at several gold and tin processing plants.

Another development, double-start spiral concentrators with two spirals integrated around a common column, have effectively doubled the capacity per unit of floor space (Fig. 4.21). At Mount Wright in Canada, 4300 double-start spirals have been used to upgrade specular hematite ore at 6900t/h at 86% recovery (Hyma and Meech, 1989). Fig. 4.22 shows an installation of double-start spirals.

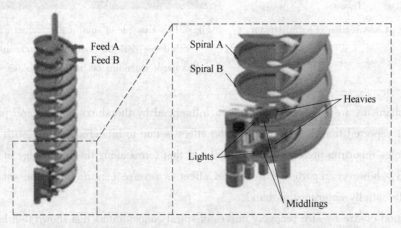

Fig. 4.21 Double-start spiral concentrator (Courtesy Multotec)

One of the most important developments in fine coal washing was the introduction in the 1980s of spiral separators specifically designed for coal. It is common practice to separate coal down to 0.5mm using dense medium cyclones (Chapter 5), and below this by froth flotation. Spiral

Fig. 4. 22 Installation of double-start spirals (Courtesy Multotec)

circuits have been installed to process the size range that is least effectively treated by these two methods, typically 0. 1~2mm (Honaker et al. ,2008).

A notable innovation in fine coal processing is the incorporation of multistage separators and circuitry, specifically recycling middling streams (Luttrell,2014). Both theoretical and field studies have shown that single-stage spirals have relatively poor separation efficiencies, as a compromise has to be made to either discard middlings or sacrifice coal yield or accept some middlings and a lower quality coal product. To address this, two-stage compound spirals have been designed in which clean coal and middlings are retreated in a second stage of spirals and the middlings from this spiral are recycled to the first spiral. This is essentially a rougher-cleaner closed-circuit configuration and it can be shown that his will give a higher separation efficiency than a single-stage separator (Section 4. 8). The separation efficiencies achievable rival those of dense medium separators.

Some of the developments in spiral technology are the result of modeling efforts. Davies et al. (1991) reviewed the development of spiral models and described the mechanism of separation and the effects of operating parameters. A semi-empirical mathematical model of the spiral has been developed by Holland-Batt (1989). Holland-Batt (1995) discussed design aspects, such as the pitch of the trough and the trough shape. A detailed CFD model of fluid flow in a spiral has been developed and validated by Matthews et al. (1998).

Spirals are made with slopes of varying steepness, the angle depending on the specific gravity of separation. Shallow angles are used, for example, to separate coal from shale, while steeper angles are used for heavy mineral-quartz separations. The steepest angles are used to separate heavy minerals from heavy waste minerals, for example, zircon (s. g. 4. 7) from kyanite and staurolite (s. g. 3. 6). Capacity ranges from 1 to 3t/h on low slope spirals to about double this for the steeper units. Spiral length is usually five or more turns for roughing duty and three turns in some cleaning units. Because treatment by spiral separators involves a multiplicity of units, the separation efficiency is very sensitive to the pulp distribution system employed. Lack of uniformity in feeding results in substantial falls in operating efficiency and can lead to severe losses in recovery. This is especially true with coalspirals (Holland-Batt,1994).

4.3.3 Shaking Tables

When a film of water flows over a flat, inclined surface, the water closest to the surface is retarded by the friction of the water absorbed on the surface and the velocity increases toward the water surface. If mineral particles are introduced into the film, small particles will not move as rapidly as large particles, since they will be submerged in the slower-moving portion of the film. Particles of high specific gravity will move more slowly than lighter particles, and so a lateral displacement of the material will be produced (Fig. 4.23).

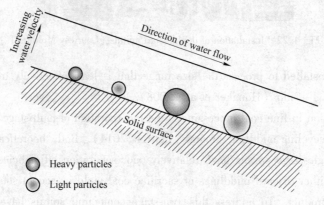

Fig. 4.23 Action in a flowing film

The flowing film effectively separates coarse light particles from small dense particles, and this mechanism is exploited to some extent in the shaking-table concentrator (Fig. 4.24), which is perhaps the most metallurgically efficient form of gravity concentrator, being used to treat small, more difficult flow-streams, and to produce finished concentrates from the products of other forms of gravity system.

The shaking table consists of a slightly inclined deck, on to which feed, at about 25% solids by weight, is introduced at the feed box and is distributed across the table by the combination of table motion and flow of water (wash water). Wash water is distributed along the length of the feed side, and the table is vibrated longitudinally, using a slow forward stroke and a rapid return, which causes the mineral particles to "crawl" along the deck parallel to the direction of motion. The minerals are thus subjected to two forces, that due to the table motion and that, at right angles to it, due to the flowing film of water. The net effect is that the particles move diagonally across the deck from the feed end and, since the effect of the flowing film depends on the size and density of the particles, they will fan out on the table, the smaller, denser particles riding highest toward the concentrate launder at the far end, while the larger lighter particles are washed into the tailings launder, which runs along the length of the table. An adjustable splitter at the concentrate end is often used to separate this product into two fractions—a high-grade concentrate (heavy product) and a middlings fraction.

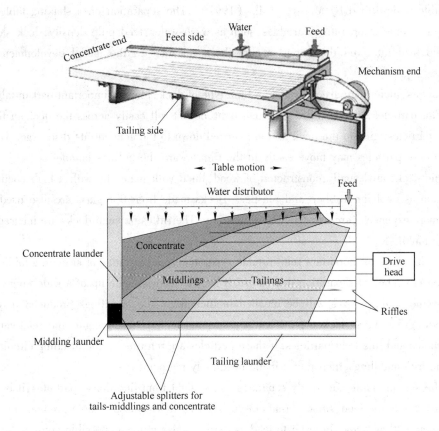

Fig. 4.24 Shaking table (concentrate refers to heavy product, etc.)

Although true flowing film concentration requires a single layer of feed, in practice, a multilayered feed is introduced onto the table, enabling much larger tonnages to be processed. Vertical stratification due to shaking action takes place behind the riffles, which generally run parallel with the long axis of the table and are tapered from a maximum height on the feed side, till they die out near the opposite side, part of which is left smooth. In the protected pockets behind the riffles, the particles stratify so that the finest and heaviest particles are at the bottom and the coarsest and lightest particles are at the top (Fig. 4.25). Layers of particles are moved across the riffles by the crowding action of new feed and by the flowing film of wash water. Due to the taper of the riffles, progressively finer sized and higher density particles are continuously being brought into contact with the flowing film of water that tops the riffles. Final concentration takes place at the unriffled area at the end of the deck, where the layer of material at this stage is usually only one or two particles deep.

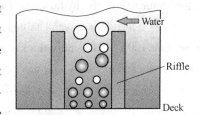

Fig. 4.25 Vertical stratification between riffles

The significance of the many design and operating variables and their interactions have been reviewed by Sivamohan and Forssberg (1985a), and the development of a mathematical model of a shaking table is described by Manser et al. (1991). The separation on a shaking table is controlled by a number of operating variables, such as wash water, feed pulp density, deck slope, amplitude, and feed rate, and the importance of these variables in the model development is discussed.

Other factors, including particle shape and the type of deck, play an important part in table separations. Flat particles, such as mica, although light, do not roll easily across the deck in the water film; such particles cling to the deck and are carried down to the concentrate discharge. Likewise, spherical dense particles may move easily in the film toward the tailings launder.

The table decks are usually constructed of wood, lined with materials with a high coefficient of friction, such as linoleum, rubber, and plastics. Decks made from fibreglass are also used which, although more expensive, are extremely hard wearing. The riffles on such decks are incorporated as part of the mold.

Particle size plays an important role in table separations; as the range of sizes in a table feed increases, the efficiency of separation decreases. If a table feed is made up of a wide range of particle sizes, some of these sizes will be cleaned inefficiently. The middlings produced may not be "true middlings", that is, locked particles of associated mineral and gangue, but relatively coarse dense particles and fine light particles. If these particles are returned to the grinding circuit, together with the true middlings, then they will be needlessly reground.

Since the shaking table effectively separates coarse light from fine dense particles, it is common practice to classify the feed, since classifiers put such particles into the same product, on the basis of their equal settling rates. In order to feed as narrow a size range as possible onto the table, classification is usually performed in multispigot hydrosizers (Chapter 3), each spigot product, comprising a narrow range of equally settling particles, being fed to a separate set of shaking tables. A typical gravity concentrator employing shaking tables may have an initial grind in rod mills to liberate as much mineral at as coarse a size as possible to aid separation, with middlings being reground before returning to the hydrosizer. Tables operating on feed sizes in the range 3mm to $100\mu m$ are sometimes referred to as sand tables, and the hydrosizer overflow, consisting primarily of particles finer than $100\mu m$, is usually thickened and then distributed to tables whose decks have a series of planes, rather than riffles, and are designated slime tables.

Dewatering of the hydrosizer overflow is often performed by hydrocyclones, which also remove particles in the overflow smaller than about $10\mu m$, which will not separate efficiently by gravity methods due to their extremely slow settling rates.

Successive stages of regrinding are a feature of many gravity concentrators. The mineral is separated at all stages in as coarse a state as possible in order to achieve reasonably fast separation and hence high throughputs.

The capacity of a table varies according to size of feed particles and the concentration criteria. Tables can handle up to 2t/h of 1.5mm sand and perhaps 1t/h of fine sand. On $100 \sim 150\mu m$ feed materials, table capacities may be as low as 0.5t/h. On coal feeds, however, which are often tabled at sizes of up to 15mm, much higher capacities are common. A normal 5mm raw coal feed can be tabled with high efficiency at 12.5t/h perdeck, while tonnages as high as 15t/h per deck are not uncommon when the feed top size is 15mm. The introduction of double and triple-deck units has improved the area capacity ratio at the expense of some flexibility and control.

Separation can be influenced by the length of stroke, which can be altered by means of a handwheel on the vibrator, or head motion, and by the reciprocating speed. The length of stroke usually varies within the range of 10~25mm or more, the speed being in the range 240~325 strokes per minute. Generally, a fine feed requires a higher speed and shorter stroke that increases in speed as it goes forward until it is jerked to a halt before being sharply reversed, allowing the particles to slide forward during most of the backward stroke due to their built-up momentum.

The quantity of water used in the feed pulp varies, but for ore-tables normal feed dilution is 20%~25% solids by weight, while for coal tables pulps of 33%~40% solids are used. In addition to the water in the feed pulp, clear water flows over the table for final concentrate cleaning. This varies from a few liters to almost 100L/min according to the nature of the feed material.

Tables slope from the feed to the tailings (light product) discharge side and the correct angle of incline is obtained by means of a handwheel. In most cases the line of separation is clearly visible on the table, so this adjustment is easily made.

The table is slightly elevated along the line of motion from the feed end to the concentrate end. The moderate slope, which the high-density particles climb more readily than the low-density minerals, greatly improves the separation, allowing much sharper cuts to be made between concentrate, middlings, and tailings. The correct amount of end elevation varies with feed size and is greatest for the coarsest and highest specific gravity feeds. The end elevation should never be less than the taper of the riffles, otherwise there is a tendency for water to flow out toward the riffle tips rather than across the riffles. Normal end elevations in ore tabling range from a maximum of 90mm for a very heavy, coarse sand, to as little as 6mm for an extremely fine feed.

Ore-concentrating tables are used primarily for the concentration of minerals of tin, iron, tungsten, tantalum, mica, barium, titanium, zirconium, and, to a lesser extent, gold, silver, thorium, and uranium. Tables are now being used in the recycling of electronic scrap to recover precious metals.

(1) Duplex Concentrator. This machine was originally developed for the recovery of tin from low-grade feeds, but has a wider application in the recovery of tungsten, tantalum, gold, chromite, and platinum from fine feeds (Pearl et al., 1991). Two decks are used alternately to provide continuous feeding, the feed slurry being fed onto one of the decks, the lower density minerals running off into the discharge launder, while the heavy minerals remain on the deck. The deck is washed with water after a preset time, in order to remove the gangue minerals, after which the deck is tilted and the concentrateis washed off. One table is always concentrating, while the other is being washed or is discharging concentrates. The concentrator has a capacity of up to 5t/h of $-100\mu m$ feed producing enrichment ratios of between 20 and 500 and is available with various sizes and numbers of decks.

(2) Mozley Laboratory Separator. This flowing film device, which uses orbital shear, is now used

in many mineral processing laboratories and is designed to treat small samples (100g) of ore, allowing a relatively unskilled operator to obtain information for a recovery grade curve within a very short time (Anon., 1980).

4.4 CENTRIFUGAL CONCENTRATORS

In an attempt to recover fine particles using gravity concentration methods, devices have been developed to make use of centrifugal force. The ability to change the apparent gravitational field is a major departure in the recovery of fine minerals.

(1) Kelsey Centrifugal Jig. The Kelsey centrifugal jig (KCJ) takes a conventional jig and spins it in a centrifuge. The main operating variables adjusted to control processing different types of feed are: centrifugal acceleration, ragging material, and feed size distribution. The 16 hutch J1800 KCJ can treat over 100t/h, depending on the application. The use of a J650 KCJ in tin recovery is described by Beniuk et al. (1994).

Other non-jig centrifugal separators have also been developed. An applied gravitational acceleration, such as that imparted by a rapidly rotating bowl, will increase the force on fine particles, allowing for easier separation based upon differences in density. Exploiting this principle, enhanced gravity, or centrifugal, concentrators were developed initially to process fine gold ores but now are applied to other minerals.

(2) Knelson Concentrator. This is a compact batch centrifugal separator with anactive fluidized bed to capture heavy minerals (Knelson, 1992; Knelson and Jones, 1994) (Fig. 4.26). A centrifugal force up to 60 times that of gravity acts on the particles, trapping denser particles in a series of rings (riffles) located in the machine, while the low-density particles are flushed out. Unit capacities range from laboratory scale to 150t/h for particles ranging in size from 10μm to amaximum of 6mm. It is generally used for feeds in which the dense component to be recovered is a very small fraction of the total material, less than 500g/t(0.05% by weight).

Feed slurry is introduced through a stationary feed tube and into the concentrate cone. When the slurry reaches the bottom of the cone it is forced outward andup the cone wall under the influence of centrifugal force. Fluidization water is introduced into the concentrate cone through a series of fluidization holes (see inset in Fig. 4.26). The slurry fills each ring to capacity to create a concentrating bed, with compaction of the bed prevented by the fluidization water. The flow of water that is injected into the rings is controlled to achieve optimum bed fluidization. High-specific gravity particles are captured and retained in the concentrating cone; the high-density material may also substitute for the low-density material that was previously in the riffles—made possible by the fluidization of the bed. When the concentrate cycle is complete, concentrates are flushed from the cone into the concentrate launder. Under normal operating conditions, this automated procedure is achieved in less than 2min in a secure environment.

The units have seen a steady improvement in design from the original Manual Discharge (MD), to Centre Discharge (CD), to Extended Duty (XD), and the Quantum Series. The first installation

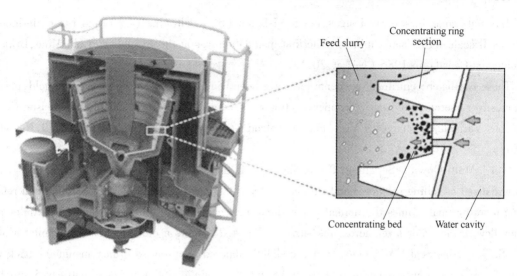

Fig. 4.26　Knelson concentrator cutaway;and action inside the riffle (Courtesy FLSmidth)

was in the grinding circuit at Camchib Mine, Chibougamau, Quebec, Canada, in 1987 (Nesset, 2011).

(3) Falcon Concentrator. Another spinning batch concentrator (Fig. 4.27), it is designed principally for the recovery of free gold in grinding circuit classifier underflows where, again, a verysmall (<1%) mass pull to concentrate is required. The feed first flows up the sides of a cone-shaped bowl, where it stratifies according to particle density before passingover a concentrate bed fluidized from behind by back-pressure (process) water. The bed retains dense particles such as gold, and lighter gangue particles are washed over the top. Periodically the feed is stopped, the bed rinsed to remove any remaining lights and is then flushed out as the heavy product. Rinsing flushing frequency, which is under automatic control, is determined from grade and recovery requirements.

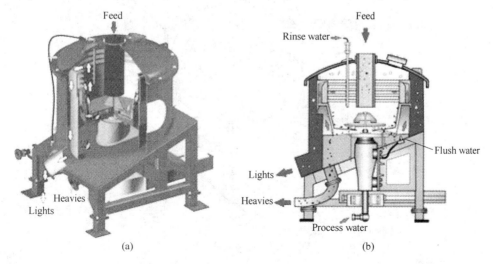

Fig. 4.27　Falcon SB Concentrator
(a) cutaway view, and (b) flow of feed and products

The units come in several designs, the Semi-Batch (SB), Ultrafine (UF), and i-Con, designed for small scaleand artisanal miners. The first installation was at the Blackdome Gold Mine, British Columbia, Canada, in 1986 (Nesset, 2011).

These two batch centrifugal concentrators have been widely applied in the recovery of gold, platinum, silver, mercury, and native copper; continuous versions are also operational, the Knelson Continuous Variable Discharge (CVD) and the Falcon Continuous (C) (Klein et al., 2010; Nesset, 2011).

(4) Multi-Gravity Separator. The principle of the multi-gravity separator (MGS) can bevisualized as rolling the horizontal surface of a conventional shaking table into a drum, then rotating it so that many times the normal gravitational pull can be exerted on the mineral particles as they flow in the water layer across the surface. Fig. 4.28 shows a cross section of the pilot scale MGS. The Mine Scale MGS consists of two slightly tapered open-ended drums, mounted "back to back", rotating at speeds variable between 90 and 150r/min, enabling forces of between 5 and 15 G to be generated at the drum surfaces. A sinusoidal shake with anamplitude variable between 4 and 6 cps is superimposed on the motion of the drum, the shake imparted to one drum being balanced by the shake imparted to the other, thus balancing the whole machine. A scraper assembly is mounted within each drum on a separate concentric shaft, driven slightly faster than the drum but in the same direction. This scrapes the settled solids up the slope of the drum, during which time they are subjected to counter-current washing before being discharged as concentrate at the open, outer, narrow end of the drum. The lower density minerals, along with the majority of the wash water, flow downstream to discharge via slots at the inner end of each drum. The MGS has been used to effect improvements in final tin concentrate grade (Turner and Hallewell, 1993).

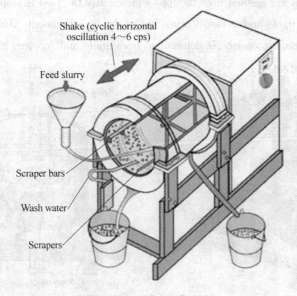

Fig. 4.28 Pilot scale MGS

(5) Testing for Gravity Recoverable Gold. Although most of the gold from gold mines worldwide

is recovered by dissolution in cyanide solution, a proportion of coarse (175μm) gold is recovered by gravity separators. It has been argued that separate treatment of the coarse gold in this way constitutes a security risk and increases costs. Gravity concentration can remain an attractive option only if it can be implemented with low capital and operating costs. A test using a laboratory centrifugal concentrator designed to characterize the gravity recoverable gold (GRG) has been described by Laplante et al. (1995), and recently reviewed by Nesset (2011). The GRG test has become a standard method for determining how much of the gold in an ore can be recovered by gravity, often through employing a centrifugal separator. The procedure for undertaking a GRG test isshown in Fig. 4.29. The sieved fractions are assayed for gold and cumulative gold recovery as a function of particle size determined (GRG response curves).

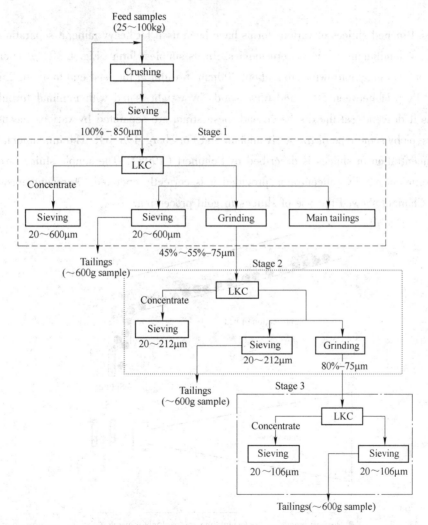

Fig. 4.29 Flowsheet detailing the GRG test (LKC is laboratory Knelson concentrator)

The results of a GRG test do not directly indicate what the gravity recovery of an installed circuit would be. The GRG test aims to quantify the ore characteristics only. In practice, plant recoveries in a gravity circuit have been found to vary from 20% to 90% of the GRG test value. The test is now applied to other high-value dense minerals such as platinum group minerals.

Often, the gravity concentrator unit will be placed inside the closed grinding circuit, treating the hydrocyclone underflow. Using GRG results, it is possible to simulate recovery by a gravity separation device, which can be used to decide on the installation and how much of the underflow to treat. In certain cases where sulfide minerals are the gravity gold carrier, flash flotation combined with gravity concentration technology provides the most effective gold recovery (Laplante and Dunne, 2002).

4.5 SLUICES AND CONES

(1) Sluices. Pinched sluices of various forms have been used for heavy mineral separations for centuries and are familiar in many western movies. In its simplest form (Fig. 4.30), it is an inclined launder about 1m long, narrowing from about 200mm in width at the feed end to about 25mm at the discharge. Pulp of between 50% and 65% solids by weight is fed with minimal turbulence and stratifies as it descends; at the discharge end these strata are separated by various means, such as by splitters, or by some type of tray (Sivamohan and Forssberg, 1985b). The fundamental basis for gravity concentration in sluices is described by Schubert (1995). The simple sluice box can be a relatively efficient gravity concentrator, provided it is correctly operated. A recent episode on the Discovery Channel showed the use of sluices in gold processing.

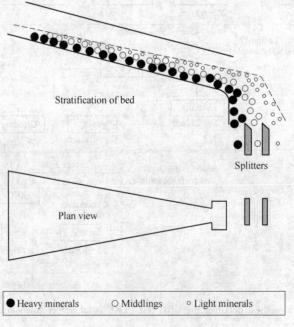

Fig. 4.30 Pinched sluice

(2) Reichert Cone. The Reichert cone is a wet gravity concentrating device that was designed for

high-capacity applications in the early 1960s, primarily to treat titanium-bearing beach sands (Ferree, 1993). Its principle of operation is similar to the pinched sluice. Fig. 4.31 shows a schematic of a Reichert cone, with the feed pulp being distributed evenly around the periphery of the cone. As it flows toward the center of the cone the heavy particles separate to the bottom of the film. A slot in the bottom of the concentrating cone removes this heavy product (usually the concentrate); the part of the film flowing over the slot is the light product (tailings). The efficiency of this separation process is relatively low and is repeated a number of times within a single machine to achieve effective performance. Inability to observe the separation is also a disadvantage (Honaker et al., 2014).

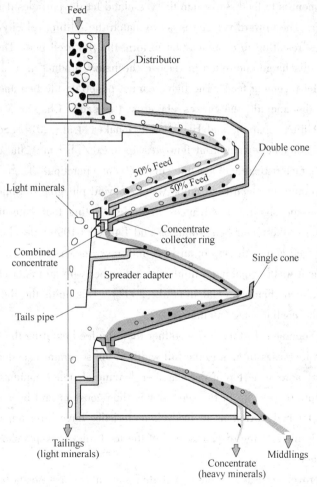

Fig. 4.31　Schematic of the Reichert cone concentrator

The success of cone circuits in the Australian mineral sand industry led to their application in other fields. At one point, Palabora Mining Co. in South Africa used 68 Reichert cones to treat 34000t/d of flotation tailings. However, the separation efficiency was always lower than spirals and as the design and efficiency of spiral concentrators improved, especially with the addition of double-start spirals, they started to retake the section of the processing industry that cones had initially gained. Today, there are relatively few circuits that include Reichert cones.

4.6 FLUIDIZED BED SEPARATORS

Fluidized bed separators (FBS), also known as teetered-bed or hindered-bed separators, have been in mineral processing plants for over a century. Initially used for size separation (Chapter 3), FBS units can be operated to provide efficient density-based separation in the particle sizerange 1 ~ 0.15mm by exploiting hindered settling and the autogeneous dense medium naturally provided by fine high-density particles in the feed.

A typical arrangement is to feed slurry into the vesseland let the particles descend against an upward current of water. The upward velocity is set to match the settling velocity of the finest fraction of the dense particles, resulting in accumulation to form the fluidized bed. The bed level is monitored, and underflow discharge controlled to remove the heavy product at a rate dependent on the mass of heavies in the incoming feed. The lights cannot penetrate the bed and report as overflow.

Units available today include the Stokes classifier (see also Chapter 3), Lewis hydrosizer, Linatex hydrosizer, Allflux separator, and Hydrosort (Honaker et al., 2014). Some positive features of FBS units include efficient separation at flowrates up to $20t/(h \cdot m^2)$, the capability toadjust to variations in feed characteristics, and general simplicity of operation. There is a need for close control of the top size and for clean fluidization water to avoid plugging the injection system.

There have been some significant advances in design. Noting that entering feed slurry could cause some disruption to the teeter bed, Mankosa and Luttrell (1999) developed a feed system to gently introduce the feed across the top of an FBS unit, now known as the Cross Flow™ Separator.

Another development was to combine a fluidization chamber with an upper chamber comprising a system of parallel inclined channels (Galvin et al., 2002). This unitis the Reflux Classifier™, now well established in the coalindustry (Bethell, 2012).

(1) Cross Flow™ Separator. In this device, rather than the feed entering the teeter bed, a tangential feed inlet, which increases in area to the full width of the separator to reduce input turbulence, directs the feed slurry across the top of the chamber, leaving chamber contents largely undisturbed (Fig. 4.32). The upward velocity in the separator is thus constant and because the feed does not directly enter the teeter bed, variations in feed characteristics have little impact on the separation performance. A baffle plate at the discharge end of the feed inlet prevents short-circuiting of solidsinto the floats product.

An additional improvement was to include a slotted plate above a series of bars carrying large diameter holes (>12.5mm) through which the fluidization water is injected. In this arrangement, the holes are simply used to introduce the water while the slotted plate acts to distribute the water, a combination that reduces the problem of plugging faced by the prior system of distribution piping.

The new design has increased separation efficiency and throughput, which combine to reduce operating costs compared to the traditional designs. Honaker et al. (2014) report that tests have been conducted on a mineral sands application and a unit has been installed in a coal plant.

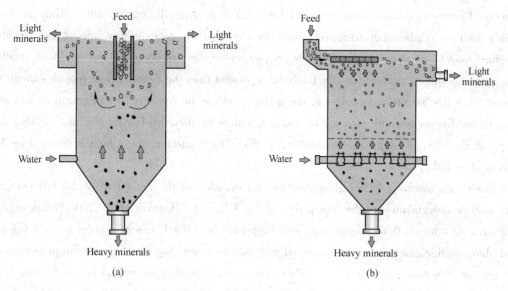

Fig. 4.32 Comparison of (a) traditional, and (b) Cross Flow separators

(2) Reflux Classifier™. The Reflux Classifier is a system of parallel inclined channels above a fluidization chamber (Galvin et al., 2002; Galvin, 2012). Using closely spaced channels promotes laminar flow (laminar-shear mechanism), which results in fine dense particles segregating and sliding downward back to the fluidization chamber while a broadsize range of light particles are transported upward to the overflow (Galvin et al., 2010). This is a version of lamella technology. An inverted version, that is, with the inlined channels at the bottom, is being developed into a flotation machine.

Ghosh et al. (2012) report on an installation of two Reflux Classifiers treating 1~0.5mm coal at 110t/h with excellent ash reduction. Pilot tests have shown potential for use in processing fine iron ore (Amarieiet al., 2014).

4.7 DRY PROCESSING

Honaker et al. (2014) review the historical development of using air as the medium to achieve gravity-based separations. The initial prime application was in coal preparation, peaking in tons treated in the United States around the mid-1960s. The pneumatic technologies follow the same basic mechanisms described for wet processing. With demand to reduce water usage growing, these technologies are worth reexamining.

(1) Pneumatic Tables. These were initially the most common pneumatic gravity devices. They use the same throwing motion as their wet counterparts to move the feed along a flat riffled deck, while blowing air continuously up through a porous bed. The stratification produced is somewhat different from that of wet tables. Whereas in wet tabling the particle size increases and the density decreases from the top of the concentrate band to the tailings, on an air table both particle size and density decrease from the top down, the coarsest particles in the middlings band having the lowest

density. Pneumatic tabling is therefore similar in effect to hydraulic classification. They are commonly used in combination with wet tables to clean zircon concentrates, one of the products obtained from heavy mineral sand deposits. Such concentrates are often contaminated with small amounts of fine quartz, which can effectively be separated from the coarse zircon particles by air tabling. Some fine zircon may be lost in the tailings and can be recovered by treatment on wet shaking tables. Recent testwork into air tabling for coal is detailed by Honaker et al. (2008) and Gupta et al. (2012). A modified air-table, the FGX Dry Separator, from China is modeled by Akbari et al. (2012).

(2) Air Jigs. Gaudin (1939) describes devices common in the early part of the last century. Two modern descendants are the Stump jig and the Allair jig (Honaker et al. ,2014). These units employ a constant air flow through a jig bed supported on a fixed screen in order to open the bed and allow stratification. Bed level is sensed and used to control the discharge rate in proportion to the amount of material to be rejected. There are several installations around the world, mostly for coal cleaning. Commercial units, for example, can treat up to 60t/h of 75~12mm coal (Honaker et al. ,2014).

(3) Other Pneumatic-Based Devices. Various units have been modified to operate dry including: the Reflux Classifier (Macpherson and Galvin,2010), Knelson concentrator (Greenwood et al. , 2013), Reichert cone (Rotich et al. ,2013), and fluidized bed devices (Franks et al. ,2013). The latter devices overlap with DMS. For example, applications in coal employ a dense medium of air and magnetite, the air dense medium fluidized bed (ADMFB) process. The first commercial ADMFB installation was a 50t/h unit in China in 1994 (Honaker et al. ,2014). Various ADMFB processes are reviewed by Sahu et al. (2009).

4.8 SINGLE-STAGE UNITS AND CIRCUITS

(1) Single Versus Two Stages of Spirals. As noted, one of the innovations in coal processing is to incorporate circuits. By making some simplifying assumptions, it is possible to deduce an analytical solution for a circuit, as Luttrell (2002) demonstrates based on the work of Meloy (1983). Fig. 4.33(a) shows two stages of spiral in a possible coal application: the rougher gives a final discard heavy product (refuse) and light product that is sent to a cleaner stage which gives the final light product (i.e., clean coal) with the heavy product (middlings) being recycled to the rougher. For this rougher-cleaner circuit the solution for the circuit recovery Rcirc is given by a mass balance across the dashed box.

Letting the feed to the rougher be X (mass units per unit time) then:

$$F = XR_c R_r + X(1 - R_r) \tag{4.5}$$

and thus circuit recovery is:

$$R_{\text{circ}} = \frac{R_c R_r}{R_c R_r + (1 - R_r)} \tag{4.6}$$

4.8 SINGLE-STAGE UNITS AND CIRCUITS · 125 ·

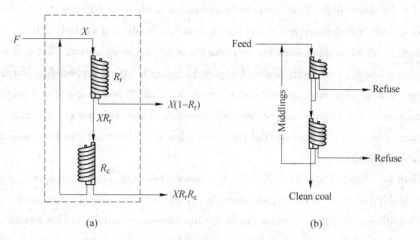

Fig. 4.33 (a) two stages of spirals in rougher-cleaner spiral combination,
and (b) a two-stage compound spiral

For each unit there is a partition curve, that is, recovery to specified concentrate as a function of particle density. Fig. 4.34 shows a generic partition curve in reduced form (Chapter 1), where the abscissae is density divided by the density corresponding to 50% recovery, that is, ρ/ρ_{50}. (Note the values of ρ/ρ_{50} decrease from left to right, indicating the concentrate is the light product, e. g., clean coal.) The slope at any point is the sharpness of the separation (the ability to separate between density classes). Assuming the same recovery point for both units, we can differentiate to solve for the sharpness of separation (slope) of the circuit, namely:

$$\frac{dR_{circ}}{dR} = \frac{2R - R^2}{(R^2 - R + 1)^2} \qquad (4.7)$$

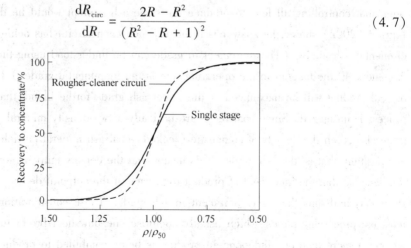

Fig. 4.34 Generic reduced form of partition curve, showing slope for
rougher-cleaner circuit with middlings recycle (Fig. 4.33(a)),
dashed line, is superior to the single unit, solid line

A convenient point for comparison is the slope at $\rho/\rho_{50} = 1$, or $R = 0.5$, which upon substituting in Eq. (4.7) gives $dR_{circ}/dR = 1.33$; in other words, the slope at $\rho/\rho_{50} = 1$ for the circuit is 1.33

times that for the single unit. This increase in sharpness is illustrated in Fig. 4. 34 (dashed line). A variety of circuits can beanalyzed using this approach (Noble and Luttrell, 2014).

The circuit, in effect, compensates for deficiencies inthe stage separation. When the stage separation efficiency ishigh, such as in DMS (Chapter 5), circuits offer less benefit. But this is rather the exception and is why circuits are widely used in mineral processing, for example, rougher-cleaner-recleaner spiral circuits in iron ore processing, and the wide range of flotation circuits is evident. Circuits are less common in coal processing, but the advantage can be demonstrated (Bethell and Arnold, 2002).

Rather than two stages of spirals, Fig. 4. 33(b) shows a two-stage compound spiral. Typically, spiral splitters will give three products, for example, in the case of a coal application: clean coal, refuse, and middlings. In the compound spiral the top three-turn spiral (in this example) produces final refuse, and the clean coal and middlings are combined and sent to the lower four-turn spiral, which produces another final refuse, final clean coal, and a middlings, which is recycled. Bethell and Arnold (2002) report that compared to two stages of spirals, the compound spiral had reduced floor space requirements with reduced capital and operating costs and was selected for a plant expansion.

(2) Parallel Circuits. Coal preparation plants usually have parallel circuits producing a final product as a blend. Luttrell (2014) describes a generic flowsheet comprising four independent circuits, each designed to treat a particular particlesize: coarse (>10mm) using dense medium baths, medium (10~1mm) using dense medium cyclones, small (1~0. 15mm) using spirals, and fine (<0. 15mm) by flotation. This arrangement poses the interesting question: what is the optimum blend of the four products to meet the target coal quality (ash grade) specification? It might seem that controlling all four to produce the same ash content would be the answer. However as Luttrell (2002) shows, the answer is to blend when each circuit has achieved the same target incremental ash grade. (The increment in grade can be understood using the grade-gradient plot in Appendix II the tangent to the operating line is the incremental grade.) In this manner the yield of coal product will be maximized at the target ash grade forthe blend; that is, in this respect the process is optimized. Since coal feeds are basically a two-density mineral situation (coal and ash minerals), then the density of composite (locked) coal-ash mineral particles will exactly reflectthe composition, that is, the ash grade. This means that the density cutpoint for each circuit should be the same so that the increment of product from each at the cutpoint density has the same increment in density and thus the same increment in ash grade. These considerations are of less import in most ore processing plants, which usually comprise one flowsheet producing final concentrate, but whenever more than one independent product is being combined to produce final concentrate this same questionof the optimized blend arises.

REFERENCES

Agricola, G. , 1556. De Re Metallica. Translated by Hoover, H. C. , and Hoover, L. H. , Dover Publications, Inc. New York, NY, USA, 1950.

Akbari, H. et al. ,2012. Application of neural network for modeling the coal cleaning performance of the FGX dry separator. In: Young, C. A. , Luttrell, G. H. (Eds.) , Separation Technologies for Minerals, Coal, and Earth Resources. SME, Englewood, Co. , USA, 189-197.

Amariei, D. et al. ,2014. The use of a Reflux Classifier for iron ores: assessment of fine particles recovery at pilot scale. Miner. Eng. ,62,66-73.

Anon, 1980. Laboratory separator modification improves recovery of coarse grained heavy minerals. Mining Mag. 142-143 (Aug.) ,158-161.

Anon,2011. Basics in Mineral Processing. Eighth ed. Metso Corporation.

Bazin, C. , et al. , 2014. Simulation of an iron ore concentration circuit using mineral size recovery curves of industrial spirals. Proceedingsof the 46th Annual Meeting of the Canadian Mineral Processors Conference, CIM , Ottawa, ON, Canada, 387-402.

Beniuk, V. G. , et al. , 1994. Centrifugal jigging of gravity concentrate and tailing at Renison Limited. Miner. Eng. ,7 (5-6) ;577-589.

Bethell, P. J. , 2012. Dealing with the challenges facing global coal preparation. In: Klima, M. S. , et al. , (Eds.) , Challenges in Fine CoalProcessing, Dewatering, and Disposal. SME, Englewood, Co. , USA, 33-45.

Bethell, P. J. Amold, B. J. 2002. Comparing a two-stage spiral to two stages of spirals for fine coal preparation. In: Honaker, R. Q. Forrest, W. R. (Eds.), Advances in Gravity Concentration. SME, Littleton, Co. , USA, 107-114.

Bonsu, A. K. , 1983. Influence of Pulp Density and Particle Size on Spiral Concentration Efficiency. Thesis (M. Phil.) , Camborne School of Mines, University of New South Wales (UNSW) , Australia.

Chen, W. L. , 1980. Batac jig cleaning in five U. S. plants. Mining Eng. 32 (9) , 1346-1350.

Cope, L. W. ,2000. Jigs: the forgotten machine. Eng. Min. J. 201 (8) ,30-34.

Davies, P. OJ, et al. 1991. Recent developments in spiral design, construction and application. Miner. Eng. 4 (3-4) ,437- 456.

Ferree, T J, 1993. Application of MDL Reichert cone and spiral concen trators for the separation of heavy minerals. CIM Bull. 86 (975) ,35-39.

Franks, G. V. , et al. ,2013. Copper ore density separations by float/sink in a dry sand fluidised bed dense medium. Int. J. Miner. Process. 121(10) ,12-20.

Galvin, K. P. 2012. Development of the Reflux Classifier. In: Klima, M. S. , et al. , (Eds.) , Challenges in Fine Coal Processing, Dewatering, and Disposal. SME, Englewood, Co. , USA, 159-186.

Galvin, K. P. , et al. ,2002. Pilot plant trial of the reflux classifier. Miner. Eng. 15 (1-2) ,19-25.

Galvin, K. P. et al. ,2010. Application of closely spaced inclined channels in gravity separation of fine particles. Miner. Eng. 23 (4) ,326-338.

Gaudin, A. M. ,1939. Principles of Mineral Dressing. McGraw-Hill Book Company, Inc. , London, England.

Ghosh, T. , et al. ,2012. Performance evaluation and optimization of a full-scale Reflux classifier. Coal Prep. Soc. Amer. J. 11 (2) ,24-33.

Gray, A. H. 1997. InLine pressure jig -an exciting, low cost technology with significant operational benefits in gravity separation of miner-als. Proceedings of the AusIMM Annual Conference, AusIMM, Ballarat, VIC, 259-266.

Green, P. , 1984. Designers improve jig efficiency. Coal Age. 89 (1) ,50-53.

Greenwood, M. , et al. 2013. The potential for dry processing using a Knelson concentrator. Miner. Eng. 45, 44-46.

Gupta, N. , et al. ,2012. Application of air table technology for cleaning Indian coal. In: Young, C. A. , Luttrell, G. H. (Eds.) , Separation Technologies for Minerals, Coal, and Earth Resources. SME, Englewood, Co. , USA, 199-210.

Holland-Batt, A. B. 1989. Spiral separation: theory and simulation. Trans. Inst. Min. Metall. Sec. C. 98 (Jan. -Apr.), C46-C60.

Holland-Batt, A. B. 1994. The effect of feed rate on the performance of coal spirals. Coal Preparation. 14 (3-4), 199-222.

Holland-Batt, A. B. ,1995. Some design considerations for spiral separators. Miner. Eng. 8 (11), 1381-1395.

Holland-Batt, A. B. ,1998. Gravity separation: a revitalized technology. Mining Eng. 50 (Sep), 43- 48.

Holland-Batt, A. B. , Holtham, P. N. ,1991. Particle and fluid motion onspiral separators. Miner. Eng. 4 (3-4), 457-482.

Honaker, R. Q. , et al. , 2008. Upgrading coal using a pneumatic densitybased separator. Int. J. Coal Prep. Utils. 28 (1), 51-67.

Honaker, R. , et al. , 2014. Density-based separation innovations in coal and minerals processing applications. In: Anderson, C. G. et al. , (Eds.) , Mineral Processing and Extractive Metallurgy: 100 Years of Innovation. SME, Englewood, Co. , USA, 243-264.

Hyma, D. B. , Meech, J. A. , 1989. Preliminary tests to improve the iron recovery from the -212 micron fraction of new spiral feed at QuebecCartier Mining Company. Miner. Eng. 2 (4), 481-488.

Jonkers, A. , et al. , 2002. Advances in modelling of stratification in jigs. Proceedings of the 13th International Coal Preparation Congress, vol. 1, Jonannesburg, South Africa, 266-276.

Klein, B. , et al. , 2010. A hybrid flotation-gravity circuit for improved metal recovery. Int. J. Miner. 'Process. 94 (3-4), 159-165.

Knelson, B. 1992. The Knelson concentrator. Metamorphosis fromcrude beginning to sophisticated world wide acceptance. Miner. Eng. 5 (10-12), 1091-1097.

Knelson, B. , Jones, R. ,1994. "A new generation of Knelson concentrators" a totally secure system goes on line. Miner. Eng. 7 (2-3), 201-207.

Kökkllhç, O. , et al. ,2015. A design of experiments investigation into dry separation using a Knelson concentrator. Miner. Eng. 72, 73-86.

Laplante, A. Dunne, R. C. , '2002. The gravity recoverable gold test and flash flotation. Proceedings of the. 34th Annual Meeting of the Canadian Mineral Processors Conference, Ottawa, ON, Canada, 105-124.

Laplante, A. R. , et al. , 1995. Predicting gravity separation gold recoveries. Miner. Metall. Process. 12 (2), 74-79.

Loveday, G. , Jonkers, A. , 2002. The Apic jig and the JigScan controller take the guesswork out of jigging. Proceedings of the 14th International Coal Preparation Congress and Exhibition, Johannesburg, South Africa, 247-251.

Luttrell, G. H. , 2002. Density separation: are we really making use of existing process engineering knowledge? In: Honaker, R. Q. , Forrest, W. R. (Eds.) , Advances in Gravity Concentration. SME, Littleton, Co. , USA, 1-16.

Luttrell, G. H. , 2014. Innovations in coal processing. In: Anderson, C. G. , et al. , (Eds.) , Mineral Processing and Extractive Metallurgy: 100Years of Innovation. SME, Englewood, Co. , USA, 277-296.

Lyman, G. J. , 1992. Review of jigging principles and control. Coal Preparation. 11 (34), 145-165.

Macpherson, S. A. , Galvin, K. , 2010. The effect of vibration on dry coal beneficiation in the Reflux classifier. Int. J. Coal Prep. Util. 30(6), 283-294.

Mankosa, M. J. , Luttrell, G. H. , 1999. Hindered-Bed Separator Device and Method. Patent No. US 6264040 B1. United States Patent Office.

Manser, R. J. , et al. , 1991. The shaking table concentrator: the influence of operating conditions and table parameters on mineral separa-. tion-the development of a mathematical model for normal operat-ing conditions. Miner.

Eng. 4 (3-4),369-381.

Matthews,B. W. ,et al. ,1998. Fluid and particulate flow on spiral concentrators: computational simulation and validation. Appl. 'Math. Model. 22 (12),965-979.

Meloy,T. P. 1983. Analysis and optimization of mineral processing and coalcleaning circuits circuit analysis. Int. J. Miner. Process. 10(1),61-80.

Miller,D. J. 1991. Design and operating experience with the Goldsworthy Mining Limited BATAC jig and spiral separator ironore beneficiation plant. Miner. Eng. 4 (3-4),411-435.

Mills,C. ,1980. Process design,scale-up and plant design for gravity concentration. In: Mular,A. L. ,Bhappu, R. B. (Eds.),MineralProcessing Plant Design,second ed. AIMME,New York,NY,404- 426. (Chapter 18).

Mishra,B. K. Mehrotra,S. P. ,1998. Modelling of particle stratification in jigs by the discrete element method. Miner. Eng. 11 (6),511-522.

Mishra,B. K. ,Mehrotra,S. P. ,2001. A jig model based on the discrete element method and its experimental validation. Int. J. Miner. Process. 63 (4),177-189.

Nesset,J. E. ,2011. Significant Canadian developments in mineral processing technology—1961 to 2011. In: Kapusta,J. ,et al. ,(Eds.),The Canadian Metallurgical & Materials Landscape 1960 to 2011. MetSoc,CIM,Westmount,Montreal,Quebec,Canada,241-293.

Noble,A. ,Luttrell,G. H. ,2014. The matrix reduction algorithm for solv-ing separation circuits. Miner. Eng. 64,97-108.

Pearl,M. ,et al. ,1991. A mathematical model of the duplex concentrator. Miner. Eng. 4 (3-4),347-354.

Priester,M. ,et al. ,1993. Tools for Mining: Techniques and Processes for Small Scale Mining. Informatica International,Incorporated (Publisher) Braunschweig,Lower Saxony,Germany.

Rotich,N. ,et al. ,2013. Modeling and simulation of gravitational solidsolid separation for optimum performance. Powder Technol. 239,337-347.

Sahu,A. K. ,et al. ,2009. Development of air dense medium fluidised bed technology for dry beneficiation of coal—a review. Int. J. CoalPrep. Util. 29 (4),216-224.

Schubert,H. ,1995. On the fundamentals of gravity concentration in sluices and spirals. Aufbereitungs Technik. 36 (11),497-505.

Sivamohan,R. ,Forssberg,E. ,1985a. Principles of tabling. Int. J. Miner. Process. 15 (4),281-295.

Sivamohan,R. ,Forssberg. E. ,1985b. Principles of sluicing. Int. J. Miner. Process. 15 (3),157-171.

Turner,J. W. G. ,Hallewell,M. P. ,1993. Process improvements for fine cassiterite recovery at Wheal Jane. Miner. Eng. 6 (8-10),817-829.

Wallace,W. M. ,1979. Electronically controlled Baum jig washing. Mine & Quarry. 8(7),43-45.

Wells,A. ,1991. Some experiences in the design and optimisation of fine gravity concentration circuits. Miner. Eng. 4 (3-4),383-398.

Xia,Y. ,et al. ,2007. CFD simulation of fine coal segregation and stratification in jigs. Int. J. Miner. Process. 82 (3),164-176.

Zhang,B. ,et al. ,2012. Development of an automatic control system for spiral concentrator,phase 1. In: Young,C. A. ,Luttrell,G. H. (Eds.),Separation Technologies for Minerals,Coal,and Earth Resources. SME,Englewood,Co. ,USA,497-508.

Zimmerman,R. E. ,1975. Performance of the Batac jig for cleaning fine and coarse coal sizes. Proceedings of the AIME Annual Meeting,AIME,Dallas,Tx. Preprint 74-F-18: 1-25.

Chapter 5 Dense Medium Separation

5.1 INTRODUCTION

Dense medium separation (DMS) is also known as heavy medium separation (HMS) or the sink-and-float process. It has two principal applications: the preconcentration of minerals, that is, the rejection of gangue prior to grinding for final liberation, and in coal preparation to produce a commercially graded end-product, that is, clean coal being separated from the heavier shale or high-ash coal. The history of the process, innovations, and failures are reviewed by Napier-Munn et al. (2014).

In principle, it is the simplest of all gravity processesand has long been a standard laboratory method for separating minerals of different specific gravity. Heavy liquids of suitable density are used, so that those minerals lessdense (lighter) than the liquid float, while those denser (heavier) than it sink (Fig. 5.1). Rather than quotinga density value, it is common to refer to specific gravity (s.g.), "relative density" (RD), or simply density.

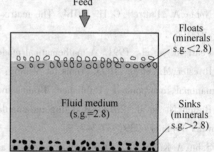

Fig. 5.1 Principle of DMS (fluid medium s.g. = 2.8 assumed for illustration)

Since most of heavy liquids are expensive or toxic, the dense medium used in industrial separations is a suspension of particles of some dense solid in water, which behaves as a heavy liquid.

The process offers some advantages over other gravity processes. It has the ability to make sharp separations at any required density, with a high degree of efficiency, even in the presence of high percentages of near-density material (or near-gravity material, i.e., material close to the desired density of separation). The density of separation can be closely controlled, within a RD of ±0.005, and can be maintained under normal conditions for indefinite periods. The separating density can be changed as required and fairly quickly, to meet varying requirements. The process is, however, rather expensive, mainly due to the ancillary equipment needed to clean and recycle the medium, and the cost of the medium itself.

For preconcentration, DMS is applicable to any ore in which, after a suitable degree of liberation by comminution, there is enough difference in specific gravity between the particles to separate those which will repay the cost of further treatment from those which will not. The process is most widely applied when the density difference occurs at a coarse particle size, for example, after crushing, as separation efficiency decreases with size due to the slower rate of settling of the particles. Particles should preferably be larger than about 4mm in diameter, in which case separation can be effective on a difference in specific gravity of 0.1 or less. If the values are finely disseminated throughout the host rock, then a suitable density difference between the crushed particles cannot be developed by coarse crushing.

Providing that a density difference exists, there is no upper size limit except that determined by the ability of the plant to handle the material. Separation down to 500μm, and less, can be facilitated by the use of centrifugal separators.

Preconcentration is most often performed on metalliferous ores that are associated with relatively light country rock, such as silicates and carbonates. Lead-zinc (galena-sphalerite) ores can be candidates, examples being the operations at Mount Isa (Queensland, Australia) and the Sullivan concentrator (British Columbia, Canada (nowclosed)). In some of the Cornish tin ores, the cassiterite is found in lodes with some degree of banded structure in which it is associated with other high-specific-gravity minerals such as the sulfides of iron, arsenic, and copper, as well as iron oxides. The lode fragments containing these minerals therefore have a greater density than the siliceous waste and allow early separation.

5.2　THE DENSE MEDIUM

5.2.1　Liquids

Heavy liquids have wide use in the laboratory for the appraisal of gravity-separation techniques on ores. Heavy liquid testing may be performed to determine the feasibility of DMS on a particular ore and to determine the economic separating density, or it may be used to assess the efficiency of an existing dense medium circuit by carrying out tests on the sink and float products. The aim is to separate the ore samples into a series of fractions according to density, establishing the relationship between the high- and the low-specific-gravity minerals (see Section 5.6).

Tetrabromoethane, having a specific gravity of 2.96, is commonly used and may be diluted with white spirit or carbon tetrachloride (s.g. 1.58) to give a range of densities below 2.96.

Bromoform (s.g. 2.89) may be mixed with carbon tetrachloride to give densities in the range 1.58~2.89. For densities up to 3.3, diiodomethane is useful, diluted as required with triethyl orthophosphate. Aqueous solutions of sodium polytungstate have certain advantages over organic liquids, such as being virtually nonvolatile, nontoxic, and of lower viscosity, and densities of up to 3.1 can easily be achieved (Plewinsky and Kamps, 1984).

For higher density separations, Clerici solution (thallium formatethallium malonate solution) allows separation at densities up to specific gravity 4.2 at 20℃ or 5.0 at 90℃. Separations of up to specific gravity 18 can be achieved by the use of magneto-hydrostatics, that is, the utilization of the supplementary weighting force produced in a solution of a paramagnetic salt or ferrofluid when situated in a magnetic field gradient. This type of separation is applicable primarily to nonmagnetic minerals with a lower limiting particle size of about 50μm (Parsonage, 1980; Domenico et al., 1994). Lin et al. (1995) describe a modification to the Franz Isodynamic Separator (Chapter 6) for use with magnetic fluids.

Many heavy liquids give off toxic fumes and must be used with adequate ventilation: the Clerici liquids are extremely poisonous and must be handled with extreme care. The use of liquids on a commercial scale has therefore not been found practicable. Magnetic fluids avoid the toxicity but attempts to use industrially also face problems of practicality, such as cleaning and recycling the expensive fluids.

For fractionating low-density materials, notably coals, solutions of salts such as calcium chloride and zinc sulfate can be used where density is controlled by concentration. Commercial application has been attempted but the problems encountered reclaiming the salts for recycle have proven difficult to surmount.

5.2.2 Suspensions

Below a concentration of about 15% by volume, finely ground suspensions in water behave essentially as simple Newtonian fluids. Above this concentration, however, the suspension becomes non-Newtonian and a certain minimum stress, or yield stress, has to be applied before shear will occur and the movement of a particle can commence. Thus, small particles, or those close to the medium density, are unable to overcome the resistance offered by the medium before movement can be achieved. This can be solved to some extent either by increasing the shearing forces on the particles or by decreasing the apparent viscosity of the suspension. The shearing force may be increased by substituting centrifugal force for gravity. The viscous effect may be decreased by agitating the medium, which causes elements of liquid to be sheared relative to each other. In practice, the medium is never static, as motion is imparted to it by paddles, air, etc., and also by the sinking material itself. All these factors, by reducing the yield stress, tend to bring the parting or separating density as close as possible to the density of the medium in the bath.

In order to produce a stable suspension of sufficiently high density, with a reasonably low viscosity, it is necessary to use fine, high-specific-gravity solid particles, agitation being necessary to maintain the suspension and to lower the apparent viscosity. The solids comprising the medium must be hard, with no tendency to slime, as degradation increases the apparent viscosity by increasing the surface area of the medium. The medium must be easily removed from the mineral surfaces by washing and must be easily recoverable from the fine-ore particles washed from the surfaces. It must not be affected by the constituents of the ore and must resist chemical attack, such as corrosion.

For ore preconcentration, galena was initially used as the medium and, when pure, it can give a bath specific gravity of about 4. Above this level, ore separation is slowed down by the viscous resistance. Froth flotation, which is an expensive process, was used to clean the contaminated medium, but the main disadvantage is that galena is fairly soft and tends to slime easily, and it also has a tendency to oxidize, which impairs the flotation efficiency.

The most widely used medium for metalliferous ores is now ferrosilicon, while magnetite is used in coal preparation. Recovery of medium in both cases is by magnetic separation. Ferrosilicon (s.g. 6.7~6.9) is an alloy of iron and silicon which should contain not less than 82% Fe and 15%-16% Si (Collins et al., 1974). If the silicon content is less than 15%, the alloy will tend to corrode, while if it is more than 16% the magnetic susceptibility and density will be greatly reduced. Losses of ferrosilicon from a dense medium circuit vary widely, from as little as 0.1 to more than 2.5kg/t of ore treated, the losses, apart from spillages, mainly occurring in magnetic separation and by the adhesion of medium to ore particles. Corrosion usually accounts for relatively small losses and can be effectively prevented by maintaining the ferrosilicon in its passive state. This is normally achieved by atmospheric oxygen diffusing into the medium or by the addition of small quantities of sodium nitrite (Stewart and Guerney, 1998).

Milled ferrosilicon is produced in a range of size distributions, from 30% to 95%-45μm, the finer grades being used for finer ores and centrifugal separators. The coarser, lower viscosity grades can achieve medium densities up to about 3.3. Atomized ferrosilicon consists of rounded particles, which produce media of lower viscosity and can be used to achieve densities up to 3.8 (Napier-Munn et al., 2014).

Magnetite (s.g. ca. 5) is used in coal washing as separation densities are not as high as needed for metalliferous ores. Medium densities are up to 2.3 but work on spheroidized magnetite aims to reach bath densities up to 2.8 (Napier-Munn et al., 2014).

5.3 SEPARATING VESSELS

Several types of separating vessel are in use, and these may be classified into gravitational ("static-baths") and centrifugal ("dynamic") vessels. There is an extensive literature on the performance of these processes, and mathematical models are being developed, which can be used for circuit design and simulation purposes (King, 2012).

5.3.1 Gravitational Vessels

Gravitational units comprise some form of vessel into which the feed and medium are introduced and the floats are removed by paddles or merely by overflow. Removal of the sinks is the most difficult part of separator design. The aim is to discharge the sinks particles without removing sufficient of the medium to cause disturbing downward currents in the vessel. They are largely restricted to treat feeds coarser than ca. 5mm in diameter. There are a wide range of gravitational devices (Leonard, 1991; Davis, 1992) and just a selection is described here.

(1) Wemco Cone Separator (Fig. 5.2). This unit is widely used for ore treatment, having a relatively high sinks capacity. The cone, which has a diameter of up to 6m, accommodates feed particles of up to 10cm in diameter, with capacities up to 500t/h.

The feed is introduced on to the surface of the medium by free-fall, which allows it to plunge several centimeters into the medium. Gentle agitation by rakes mounted on the central shaft (stirring mechanism) helps keep the medium in suspension. The float fraction simply overflows a weir, while the sinks are removed by pump (Fig. 5.2(a)) or by external or internal air lift (Fig. 5.2(b)).

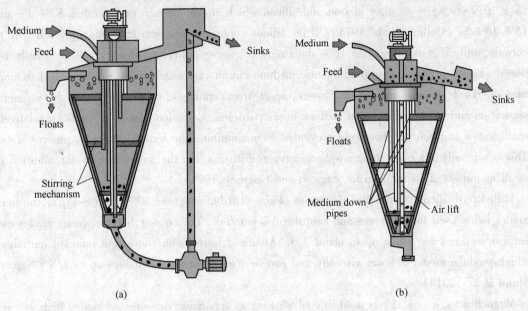

Fig. 5.2 Wemco cone separator
(a) with torque-flow-pump sinks removal, and (b) with compressed-air sinks removal

(2) Drum Separators (Fig. 5.3). These are built in several sizes, up to 4.3m diameter by 6m long, with capacities of up to 450t/h and treating feed particles of up to 30cm in diameter. Separation is accomplished by the continuous removal of the sink product through the action of lifters fixed to the inside of the rotating drum. The lifters empty into the sinks launder when they pass the horizontal position. The float product overflows a weir at the opposite end of the drum from the feed chute. Longitudinal partitions separate the float surface from the sink-discharge action of the revolving lifters.

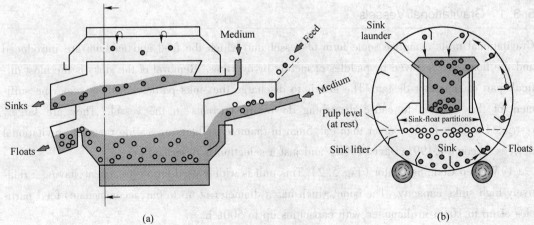

Fig. 5.3 Dense media (DM) drum separator
(a) side view, and (b) end view

The comparatively shallow pool depth in the drum compared with the cone separator minimizes settling out of the medium particles, giving a uniform gravity throughout the drum.

Where single-stage dense-medium treatment is unable to produce the desired recovery, two-stage separation can be achieved in the two-compartment drum separator (Fig. 5.4), which is, in effect, two drum separators mounted integrally and rotating together, one feeding the other. The lighter medium in the first compartment separates a true float product. The sink product is lifted and conveyed into the second compartment, where the middlings and the true sinks are separated.

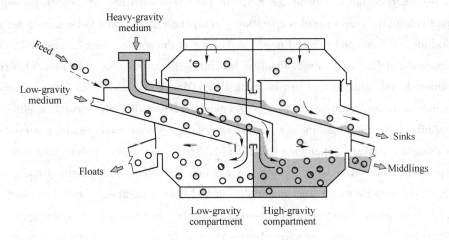

Fig. 5.4 Two-compartment DM drum separator

Although drum separators have large sinks capacities and are inherently more suited to the treatment of metallic ores, where the sinks product is normally 60% ~ 80% of the feed, they are common in coal processing, where the sinks product is only 5% ~ 20% of the feed, because of their simplicity, reliability, and relatively small maintenance needs. A mathematical model of the DM drum has been developed by Baguley and Napier-Munn (1996).

(3) Drewboy Bath. Once widely employed in the UK coal industry because of its high floats capacity, it is still in use (Cebeci and Ulusoy, 2013). The raw coal is fed into the separator at one end, and the floats are discharged from the opposite end by a star-wheel with suspended rubber, or chain straps, while the sinks are lifted out from the bottom of the bath by a radial-vaned wheel mounted on an inclined shaft. The medium is fed into the bath at two points—at the bottom of the vessel and with the raw coal—the proportion being controlled by valves.

(4) Norwalt Washer. Developed in South Africa, most installations are still to be found in that country. Raw coal is introduced into the center of the annular separating vessel, which is provided with stirring arms. The floats are carried round by the stirrers and are discharged over a weir on the other side of the vessel, being carried out of the vessel by the medium flow. The heavies sink to the bottom of the vessel and are moved along by scrapers attached to the bottom of the stirring arms and are discharged via a hole in the bottom of the bath into a sealed elevator, either of the wheel or bucket type, which continuously removes the sinks product.

5.3.2 Centrifugal Separators

Cyclonic dense medium separators have now become widely used in the treatment of ores and coal. They provide a high centrifugal force and a low viscosity in the medium, enabling much finer separations to be achieved than in gravitational separators. Feed to these devices is typically deslimed at about 0.5mm, to avoid contamination of the medium with slimes and to simplify medium recovery. A finer medium is required than with gravitational vessels, to avoid fluid instability. Much work has been carried out to extend the range of particle size treated by centrifugal separators. This is particularly the case in coal preparation plants, where advantages to be gained are elimination of desliming screens and reduced need for flotation of the screen undersize, as well as moreaccurate separation of fine coal. Froth flotation has little effect on sulfur reduction, whereas pyrite can be removed, and oxidized coal can be treated by DMS. Work has shown that good separations can be achieved for coal particles as fine as 0.1mm, but below this size separation efficiency decreases rapidly. Tests on a leadzinc ore have shown that good separations can be achieved down to 0.16mm using a centrifugal separator (Ruff, 1984). These, and similar results elsewhere, together with the progress made in automatic control of medium consistency, add to the growing evidence that DMS can be considered for finer material than had been thought economical or practical until recently. As the energy requirement for grinding, flotation, and dewatering is often up to 10 times that required for DMS, a steady increase of fines preconcentration DMS plants is likely.

(1) Dense Medium Cyclones (DMC). By far the most widely used centrifugal DM separator is the cyclone (DMC) (Fig. 5.5) whose principle of operation is similar to that of the conventional hydrocyclone (Chapter 3). Cyclones typically treat ores and coal in the range 0.5~40mm. Cyclones up to 1m in diameter for coal preparation were introduced in the 1990s, and units up to 1.4m diameter and capable of throughputs of over 250t/h treating feed particles up to 75~90mm are now common in the coal industry (Luttrell, 2014). Osborne (2010) has documented the

Fig. 5.5 Cast iron dense medium cyclones used in the coal and diamond industry
(Courtesy Multotec)

decrease in circuit complexity that has accompanied this increase in unit size. The larger DMC units treating coarser sizes may obviate the requirement for static-bath vessels, and the need for fewer units minimizes differences in cut-point densities and surges that are common in banks of smaller units.

DMC sizes have lagged in metalliferous operations, the largest being 0.8m diameter, but confer similar advantages as experience at Glencore's leadzinc plant at Mount Isa has shown (Napier-Munn et al.,2009).

The feed is suspended in the medium and introduced tangentially to the cyclone either via a pump or it is gravity-fed. Gravity feeding requires a taller and therefore more expensive building, but achieves a more consistent flow and less pump wear and feed degradation. The dense material (reject in the case of coal, product in the case of iron ore, for example) is centrifuged to the cyclone wall and exits at the apex. The light product "floats" to the vertical flow around the axis and exits via the vortex finder. In a DMC, there is a difference in density at various points. Fig. 5.6 shows a rough indication of the density variations in a DMC containing medium only (Bekker,2014). The figure is constructed assuming a density cutpoint at a RD (RD_{50}) of 2.85 (RD_{50} refers the density of a particle that has a 50:50 chance of reporting to either floats or sinks, see later). Fig. 5.6 is an idealized representation, as in reality there are density gradients radially across the cyclone as well, which the mathematical and computational models of DMCs are showing. Mathematical models of the DMC for coal were developed by King and Juckes (1988); and for minerals by Scott and Napier-Munn (1992). More recently, computational fluid dynamics models of DMCs have been developed, revealing further detail on the flows inside the device (Kuang et al.,2014).

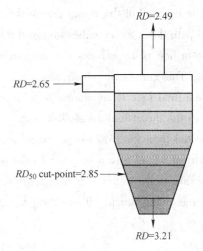

Fig. 5.6 Density gradients inside a dense medium cyclone (Adapted from Bekker,2014)

In general, DMCs have a cone angle of 20°, with manufacturers generally staying with one type of cone angle as there has been shown to be no real benefit achieved by altering it (Bekker, 2014).

(2) Water-Only Cyclones. Particles below ca. 0.5~1mm are generally too fine for the drainage and washing screens used as part of the circuit to recover/recycle DM (see Section 5.4), and particles in the range 0.2–1mm are therefore processed by water-based gravity techniques. In the coal industry, the most common such device is the spiral concentrator (Chapter 4) but water-only cyclones are also used (Luttrell,2014). They separate coal from rock within a self-generated (autogenous) dense medium derived from fine fraction of the heavy minerals in the feed (similar in concept to the fluidized bed separators, Chapter 4). Modern units have a wide angle conical bottom to emphasize density separation and suppress size effects.

(3) Vorsyl Separator (Fig. 5.7). Developed in the 1960s at the British Coal Mining Research and Development Establishment for processing 50~5mm sized feeds at up to 120t/h, the unit continues to be used (Banerjee et al., 2003; Majumder et al., 2009). The feed to the separator, consisting of de-slimed raw coal, together with the separating medium of magnetite, is introduced tangentially, or more recently by an involute entry (see Chapter 3), at the top of the separating chamber, under pressure. Material of specific gravity less than that of the medium passes into the clean coal outlet via the vortex finder, while the near-density material and the heavier shale particles move to the wall of the vessel due to the centrifugal acceleration induced. The particles move in a spiral path down the chamber toward the base of the vessel where the drag caused by the proximity of the orifice plate reduces the tangential velocity and creates a strong inward flow toward the throat. This carries the shale, and near-density material, through zones of high centrifugal force, where a final precise separation is achieved. The shale, and a proportion of the medium, discharge through the throat into the shallow shale chamber, which is provided with a tangential outlet, and is connected by a short duct to a second shallow chamber known as the vortextractor. This is also a cylindrical vessel with a tangential inlet for the medium and reject and an axial outlet. An inward spiral flow to the outlet is induced, which dissipates the inlet pressure energy and permits the use of a large outlet nozzle without the passing of an excessive quantity of medium.

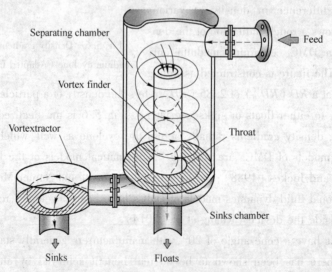

Fig. 5.7 Vorsyl separator

(4) LARCODEMS (Large Coal Dense Medium Separator). This was developed to treat a wide size range of coal (-100mm) at high capacity in one vessel (Shah, 1987). The unit (Fig. 5.8) consists of a cylindrical chamber which is inclined at approximately 30° to the horizontal. Medium at the required RD is introduced under pressure, either by pump or static head, into the involute tangential inlet at the lower end. At the top end of the vessel is another involute tangential outlet connected to a vortextractor. Raw coal of 0.5~100mm is fed into the separator by a chute connected to the top end, the clean coal being removed through the bottom outlet. High RD particles

pass rapidly to the separator wall and are removed through the top involute outlet and the vortex-tractor.

The first installation of the device was in the 250t/h coal preparation plant at Point of Ayr Colliery in the United Kingdom (Lane,1987). In addition to coal processing, the LARCODEMS has found application in concentrating iron ore, for example, a 1. 2m LARCODEMS is used in Kumba's iron ore concentrator at Sishen in South Africa to treat up to 800t/h of -90~+6mm feeds (Napier-Munn et al. ,2014) ,and in recycling, notably of plastics (Pascoe and Hou,1999;Richard et al. , 2011).

(5)Dyna Whirlpool Separator (Fig. 5. 9). Developed in the United States, this device is similar to the LARCODEMS and is used for treating fine coal, particularly in the Southern Hemisphere, as well as diamonds, fluorspar, tin, and leadzinc ores, in the size range 0. 5~30mm (Wills and Lewis, 1980).

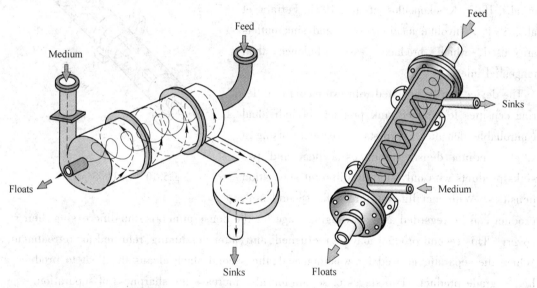

Fig. 5. 8　LARCODEMS separator　　　　　Fig. 5. 9　Dyna Whirlpool Separator

It consists of a cylinder of predetermined length having identical tangential inlet and outlet sections at either end. The unit is operated in an inclined position and medium of the required density is pumped under pressure into the lower outlet. The rotating medium creates a vortex throughout the length of the unit and leaves via the upper tangential discharge and the lower vortex outlet tube. Raw feed entering the upper vortex tube is sluiced into the unit by a small quantity of medium and a rotational motion is quickly imparted by the open vortex. Float material passes down the vortex and does not contact the outer walls of the unit, thus greatly reducing wear. The floats are discharged from the lower vortex outlet tube. The heavy particles (sinks) of the feed penetrate the rising medium toward the outer wall of the unit and are discharged with medium through the sink discharge pipe. Since the sinks discharge is close to the feed inlet, the sinks are removed from the unit almost immediately, again reducing wear considerably. Only neardensity particles, which are separated further along the unit, actually come into contact with the main cylindrical body. The tangential sink discharge outlet is connected to a flexible sink hose and the height of this hose may be used to adjust back pressure to finely control the cutpoint.

The capacity of the separator can be as high as 100t/h, and it has some advantages over the DM cyclone. Apart from the reduced wear, which not only decreases maintenance costs but also maintains performance of the unit, operating costs are lower, since only the medium is pumped. The unit has a higher sinks capacity and can accept large fluctuations in sink/float ratios (Hacioglu and Turner, 1985).

(6) Tri-Flo Separator (Fig. 5. 10). This can be regarded as two Dyna Whirlpool separators joined in series and has been installed in a number of coal, metalliferous, and nonmetallic ore treatment plants (Burton et al., 1991; Kitsikopoulos et al., 1991; Ferrara et al., 1994). Involute medium inlets and sink outlets are used, which produce less turbulence than tangential inlets.

The device can be operated with two media of differing densities to produce sink products of individual controllable densities. Two-stage treatment using a single medium density produces a float and two sinks products with only slightly different separation densities. With metalliferous ores, the second sink product can be regarded as a scavenging stage for the dense minerals, thus increasing their recovery. This second product may be recrushed, and, after de-sliming, returned for retreatment. Where the separator is used for washing coal, the second stage cleans the float to produce a higher grade product. Two stages of separation also increase the sharpness of separation.

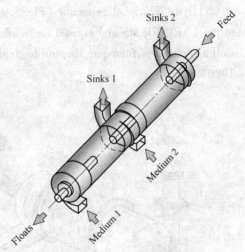

Fig. 5. 10 Tri-Flo Separator

5.4 DMS CIRCUITS

Although the separating vessel is the most important element of a DMS process, it is only one part of a relatively complex circuit. Other equipment is required to prepare the feed, and to recover, clean, and recirculate the medium (Symonds and Malbon, 2002).

The feed to a dense medium circuit must be screened to remove fines, and slimes should be removed by washing, thus alleviating any tendency that such slime content may have for creating sharp increases in medium viscosity.

The greatest expense in any dense medium circuit is for reclaiming and cleaning the medium, which leaves the separator with the sink and float products. A typical circuit is shown in Fig. 5. 11.

The sink and float fractions pass onto separate vibrating drainage screens, where more than 90%

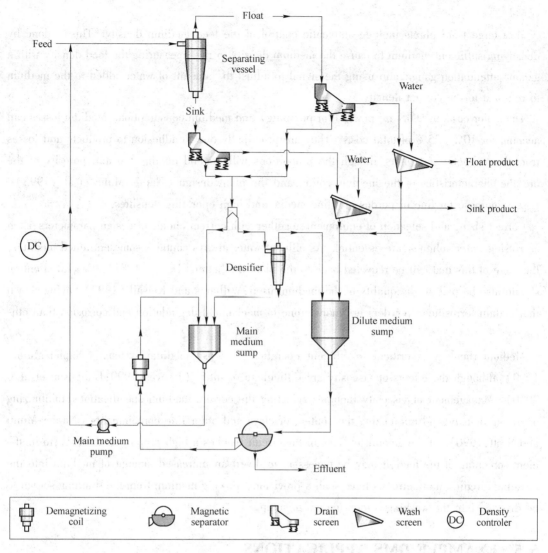

Fig. 5.11 Typical DMS circuit

of the medium in the separator products is recovered and pumped back via a sump into the separating vessel. The products then pass to wash screens, where washing sprays substantially complete the removal of medium and adhering fines. The finished float and sink (screen overflow) products are discharged from the screens for disposal or further treatment.

The underflows from the drainage screens are combined and a fraction reports to the main medium sump and the remainder is densified by a centrifugal or spiral densifier. The underflows from washing screens, consisting of medium, wash water, and fines, are too dilute and contaminated to be returned directly as medium to the separating vessel. They are treated (together in this case) by magnetic separation to recover the magnetic ferrosilicon or magnetite from the nonmagnetic fines, which also densifies the medium. The densified medium is directed to the main medium sump passing via a demagnetizing coil to ensure a nonflocculated, uniform suspension in the separating vessel.

Most large DMS plants include automatic control of the feed medium density. This is done by densifying sufficient medium to cause the medium density to rise, measuring the feed density with a gamma attenuation gauge, and using the signal to adjust the amount of water added to the medium to return it to the correct density.

The major costs in DMS are power (for pumping) and medium consumption. Medium losses can account for 10%~35% of total costs. They are principally due to adhesion to products and losses from the magnetic separators, though the proportions will depend on the size and porosity of the ore, the characteristics of the medium solids, and the plant design (Napier-Munn et al., 1995). Losses increase for fine or porous ore, fine media, and high operating densities.

Correct sizing and selection of equipment, together with correct choice of design parameters, such as rinsing water volumes, are essential. As effluent water always contains some entrained medium, the more of this that can be recycled back to the plant the better (Dardis, 1987). Careful attention should also be paid to the quality of the medium used, Williams and Kelsall (1992) having shown that certain ferrosilicon powders are more prone to mechanical degradation and corrosion than others.

Medium rheology is critical to efficient operation of dense medium systems (Napier-Munn, 1990), although the effects of viscosity are difficult to quantify (Reeves, 1990; Dunglison et al., 2000). Management of viscosity includes selecting the correct medium specifications, minimizing operating density, and minimizing the content of clays and other fine contaminants (Napier-Munn and Scott, 1990). If the amount of fines in the circuit reaches a high proportion due, say, to inefficient screening of the feed, it may be necessary to divert an increased amount of medium into the cleaning circuit. Many circuits have such a provision, allowing medium from the draining screen to be diverted into the washing screen undersize sump.

5.5 EXAMPLE DMS APPLICATIONS

The most important use of DMS is in coal preparation, where a relatively simple separation removes the low-ash coal (clean coal) from the heavier high-ash discard and associated shales and sandstones. DMS is preferred to the cheaper jigs when washing coals with a relatively large proportion of middlings, or near-density material, since the separating density can be controlled at much closer limits.

Luttrell (2014) gives a generic flowsheet for a modern US coal processing plant with four independent circuits treating different size fractions: coarse size (+10mm) using dense medium (i.e., static) vessels; medium size (-10+1mm) using dense medium cyclones; small size (-1 +0.15mm) using spirals; and fine size (-0.15mm) using flotation. British coals, in general, are relatively easy to wash, and jigs are used in many cases. Where DMS is preferred, drum and Drewboys separators are most widely used for the coarser fractions, with DM cyclones and Vorsyl separators being preferred for the fines. DMS is essential with most Southern Hemisphere coals, where a high middlings fraction is present. This is especially so with the large, low-grade coal deposits found in the former South African Transvaal province. Drums and Norwalt baths are the most common separators utilized to wash such coals, with DM cyclones and Dyna Whirlpools being used to treat the finer fractions.

At the Landau Colliery in the Transvaal (operated by Anglo Coal), a two-density operation is carried out to produce two saleable products. After preliminary screening of the run-of-mine coal, thecourse (+7mm) fraction is washed in Norwalt bath separators, utilizing magnetite as the medium to give a separating density of 1.6. The sinks product from this operation, consisting predominantly of sand and shales, is discarded, and the floats product is routed to Norwalt baths operating at a lower density of 1.4. This separation stage produces a low-ash floats product, containing about 7.5% ash, which is used for metallurgical coke production, and a sinks product, which is the process middlings, containing about 15% ash, which is used as power-station fuel. The fine (0.5–7mm) fraction is treated in a similar two-stage manner using Dyna Whirlpool separators.

In metalliferous mining, DMS is used in the preconcentration of leadzinc ores, in which the disseminated sulfide minerals often associate together as replacement bandings in the light country rock, such that marked specific gravity differences between particles crushed to fairly coarse sizes can be exploited.

A dense medium plant was incorporated into the leadzinc circuit at Mount Isa Mines Ltd., Australia, in 1982 in order to increase the plant throughput by 50%. The ore, containing roughly 6.5% lead, 6.5% zinc, and 200ppm silver, consists of galena, sphalerite, pyrite, and other sulfides finely disseminated in distinct bands in quartz and dolomite. Liberation of the ore into particles which are either sulfide-rich or predominantly gangue begins at around −50mm and becomes substantial below 18mm.

The plant treats about 800t/h of material, in the size range 1.7~13mm by DM cyclones, at a separating RD of 3.05, to reject 30%~35% of the run-of-mine ore as tailings, with 96%~97% recoveries of lead, zinc, and silver to the preconcentrate. The preconcentrate has a 25% lower Bond Work Index and is less abrasive because the lower specific gravity hard siliceous material mostly reports to the rejects. (The grinding circuit product (cyclone over flow) will also be finer after installation of DMS simply due to the removal of the low-specific-gravity mineral fraction that classifies at a coarser size, as discussed in Chapter 3.) The rejects are used as a cheap source of fill for underground operations. The plant is extensively instrumented, the process control strategy being described by Munro et al. (1982).

DMS is also used to preconcentrate tin and tungsten ores, and nonmetallic ores such as fluorite and barite. It is important in the preconcentration of diamond ores, prior to recovery of the diamonds by electronic sorting (Chapter 7) or grease-tabling (Rylatt and Popplewell, 1999). Diamonds are the lowest grade of all ores mined, and concentration ratios of several million to one must be achieved. DMS produces an initial enrichment of the ore in the order of 100 ~ 1000 to 1 by making use of the fact that diamonds have a fairly high specific gravity (3.5) and are relatively easily liberated from the ore, since they are loosely held in the parent rock. Gravitational and centrifugal separators are utilized, with ferrosilicon as the medium, and separating densities between 2.6 and 3.0. Clays in the ore sometimes present a problem by increasing the medium viscosity, thus reducing separating efficiency and the recovery of diamonds to the sinks.

Upgrading low-grade iron ores for blast furnace feed sometimes uses DMS. Both gravity and centrifugal separators are employed, and in some cases the medium density can exceed 4 (Myburgh, 2002).

5.6 LABORATORY HEAVY LIQUID TESTS

Laboratory testing may be performed on ores to assess the suitability of DMS (and other gravity methods) and to determine the economic separating density.

Liquids covering a range of densities in incremental steps are prepared, and the representative sample of crushed ore is introduced into the liquid of highest density. The floats product is removed and washed and placed in the liquid of next lower density, whose float product is then transferred to the next lower density and so on. The sinks product is finally drained, washed, and dried, and then weighed, together with the final floats product, to give the density distribution of the sample by weight (Fig. 5.12).

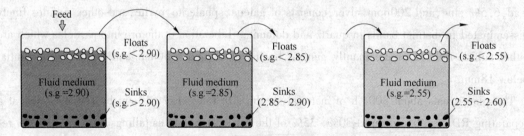

Fig. 5.12 Heavy liquid testing

Care should be taken when evaluating ores of fine particle size that sufficient time is given for the particles to settle into the appropriate fraction. Centrifuging is often carried out on fine materials to reduce the settling time, but this should be done carefully, as there is a tendency for the floats to become entrained in the sinks fraction. Unsatisfactory results are often obtained with porous materials, such as magnesite ores, due to the entrainment of liquid in the pores, which changes the apparent density of the particles.

After assaying the fractions for metal content, the distribution of material and metal in the density fractions of the sample can be tabulated. Table 5.1 shows such a distribution from tests performed on a tin ore. The computations are easily accomplished in a spreadsheet. Columns 2, 3, and 4 are self-explanatory. Column 5 is an intermediate calculation step, referred to as "Sn units", and is the product of $w\%$ and $w(\mathrm{Sn})\%$ (i. e., column 2 ×column 4). Column 6, $w(\mathrm{Sn})\%$ Distribution, is then computed by dividing each row in column 5 by the sum of units, 111. 90; column 7 is then obtained by cumulating the rows in column 6. Knowing the sum of units gives a back-calculated value of the feed (or head) assay, in this case 1. 12% Sn (i. e., sum of units divided by sum of increment weights, 111. 90/100 as a percent).

It can be seen from columns 3 and 7 of the table that if a separation density of 2. 75 was chosen, then 68. 48% of the material, being lighter than 2. 75, would be discarded as a float product, and only 3. 78% of the tin would be lost in this fraction. Conversely, 96. 22% of the tin would be recovered into the sink product (i. e., 100~3. 78), which accounts for 31. 52% of the original total feed weight. From this information we can quickly calculate the grade of Sn in the sinks by using the definition of recovery (Chapter 1):

$$R = \frac{Cc}{Ff} \tag{5.1}$$

where $R=96.22\%$, C/F (weight recovery, or yield) = 31. 52%, and f = 1. 12%, hence, solving for c, the tin grade in the sinks product, we find c = 3. 41% Sn. The analogous calculation can be used to determine the Sn grade in the discard (light) product, which gives t = 0. 062%.

The choice of optimum separating density must be made on economic grounds. In the example shown in Table 5. 1, the economic impact of rejecting 68. 48% of the feed to DMS on downstream performance must be assessed. The smaller throughput will lower grinding and concentration operating costs, the impact on grinding energy and steel costs often being particularly high. Against these savings, the cost of operating the DMS plant and the impact of losing 3. 78% of the run-of-mine tin to floats must be considered. The amount of recoverable tin in this fraction has to be estimated, together with the subsequent loss in smelter revenue. If this loss is lower than the saving in overall milling costs, then DMS is economic. The optimum density is that which maximizes the difference between overall reduction in milling costs per ton of run-of-mine ore and loss in smelter revenue. Schena et al. (1990) have analyzed the economic choice of separating density.

Heavy liquid tests are important in coal preparation to determine the required density of separation and the expected yield of coal of the required ash content. The "ash" content refers to the amount of incombustible material in the coal. Since coal is lighter than the contained minerals, the higher the density of separation the higher is the yield (Chapter 1):

$$\mathrm{yield} = \frac{\text{weight of coal floats product} \times 100\%}{\text{total feed weight}} \tag{5.2}$$

but the higher is the ash content. The ash content of each density fraction from heavy liquid testing is determined by taking about 1 g of the fraction, placing it in a cold well-ventilated furnace, and slowly raising the temperature to 815℃, maintaining the sample at this temperature until constant weight is obtained. The residue is cooled and then weighed. The ash content is the mass of ash expressed as a percentage of the initial sample weight taken.

Table 5.1　Heavy Liquid Test Results on Tin Ore Sample

1	2	3	4	5	6	7
Specific Gravity Fraction	Weight/%		$w(Sn)$% in s.g. fraction	Sn units	$w(Sn)$/%	
	Incremental	Cumulative			Distribution	Cum. Distribution
<2.55	1.57	1.57	0.003	0.0047	0.004	0.004
2.55~2.60	9.22	10.79	0.04	0.37	0.33	0.33
2.60~2.65	26.11	36.90	0.04	1.04	0.93	1.27
2.65~2.70	19.67	56.57	0.04	0.79	0.70	1.97
2.70~2.75	11.91	68.48	0.17	2.02	1.81	3.78
2.75~2.80	10.92	79.40	0.34	3.71	3.32	7.10
2.80~2.85	7.87	87.27	0.37	2.91	2.60	9.70
2.85~2.90	2.55	89.82	1.30	3.32	2.96	12.66
>2.90	10.18	100.00	9.60	97.73	87.34	100.00
Total			1.12	111.90		

Table 5.2 shows the results of heavy liquid tests performed on a coal sample. The coal was separated into the density fractions shown in column 1, and the weight fractions and ash contents are tabulated in columns 2 and 3, respectively. The weight percent of each product is multiplied by the ash content to give the ash units (column 4) (same calculation as "units" in Table 5.1).

The total floats and sinks products at the various separating densities shown in column 5 are tabulated in columns 6-11. To obtain the cumulative floats at each separation density, columns 2 and 4 are cumulated from top to bottom to give columns 6 and 7, respectively. Column 7 is then divided by column 6 to obtain the cumulative percent ash (column 8). Cumulative sink ash is obtained in essentially the same manner, except that columns 2 and 4 are cumulated from bottom to top to give columns 9 and 10, respectively. The results are plotted in Fig. 5.13 as typical washability curves.

Suppose an ash content of 12% is required in the coal product. It can be seen from the washability curves that such a coal would be produced at a yield of 55% (cumulative percent floats), and the required density of separation is 1.465.

The difficulty of the separation in terms of operational control is dependent mainly on the amount of material present in the feed that is close to the required density of separation. For instance, if the feed were composed entirely of pure coal at specific gravity 1.3 and shale at specific gravity 2.7, then the separation would be easily carried out over a wide range of operating densities. If, however, the feed consists of appreciable middlings, and much material present is near-density (i.e., very close to the chosen separating density), then only a small variation in this density will seriously affect the yield and ash content of the product.

5.6 LABORATORY HEAVY LIQUID TESTS

Table 5.2 Heavy Liquid Test Results on a Coal Sample

1	2	3	4	5	6	7	8	9	10	11
Sp. gr. fraction	w/%	Ash/%	Ash units	Separating density	Cumulative float			Cumulative sink		
					w/%	Ash units	Ash/%	w/%	Ash units	Ash/%
<1.30	0.77	4.4	3.39	1.30	0.77	3.39	4.4	99.23	2213.76	22.3
1.30~1.32	0.73	5.6	4.09	1.32	1.50	7.48	5.0	98.50	2209.67	22.4
1.32~1.34	1.26	6.5	8.19	1.34	2.76	15.67	5.7	97.24	2201.48	22.6
1.34~1.36	4.01	7.2	28.87	1.36	6.77	44.54	6.6	93.23	2172.61	23.3
1.36~1.38	8.92	9.2	82.06	1.38	15.69	126.60	8.1	84.31	2090.55	24.8
1.38~1.40	10.33	11.0	113.63	1.40	26.02	240.23	9.2	73.98	1976.92	26.7
1.40~1.42	9.28	12.1	112.29	1.42	35.30	352.52	10.0	64.70	1864.63	28.8
1.42~1.44	9.00	14.1	126.90	1.44	44.30	479.42	10.8	55.70	1737.73	31.2
1.44~1.46	8.58	16.0	137.28	1.46	52.88	616.70	11.7	47.12	1600.45	34.0
1.46~1.48	7.79	17.9	139.44	1.48	60.67	756.14	12.5	39.33	1461.01	37.1
1.48~1.50	6.42	21.5	138.03	1.50	67.09	894.17	13.3	32.91	1322.98	40.2
>1.50	32.91	40.2	1322.98	—	100.00	2217.15	22.2	—	—	—
Total	100.0	22.2	2217.15							

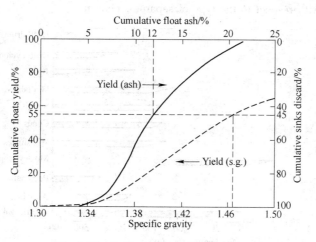

Fig. 5.13 Typical coal washability curves

The amount of near-density material present is sometimes regarded as being the weight of material in the range ±0.1 or ±0.05 of the separating RD. Separations involving feeds with less than about 7% of ±0.1 neardensity material are regarded by coal preparation engineers as being fairly easy to control. Such separations are often performed in Baum jigs, as these are cheaper than dense medium plants, which require expensive mediacleaning facilities, and no feed preparation (i.e., removal of the fine particles by screening) is required. However, the density of separation in jigs is not as easy to control to fine limits, as it is in DMS, and for near-density material much above 7%, DMS is preferred.

Heavy liquid tests can be used to evaluate any ore, and combined with Table 5.3 can be used to indicate the type of separator that could effect the separation in practice (Mills, 1980).

Table 5.3 Gravity Separation Process Depends on Amount of Near-Density Material

wt% within±0.1 RD of separation	Gravity process recommended	Type
0~7	Almost any process	Jigs, tables, spirals
7~10	Efficient process	Sluices, cones, DMS
10~15	Efficient process with good operation	
15~25	Very efficient process with expert operation	DMS
>25	Limited to few exceptionally efficient processes with expert operation	DMS with close control

Table 5.3 takes no account of the particle size of the material and experience is therefore required in its application to heavy liquid results, although some idea of the effective particle size range of gravity separators can be gained from Fig. 5.14. The throughput of the plant must also be taken into account with respect to the type of separator chosen.

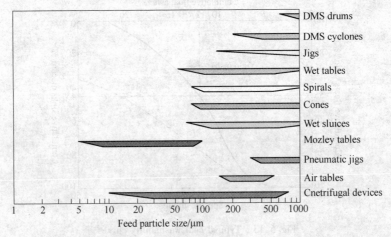

Fig. 5.14 Effective range of gravity and dense medium devices (Adapted from Mills (1980))

5.7 EFFICIENCY OF DMS

Laboratory testing assumes perfect separation and, in such batch tests, conditions are indeed close to the ideal, as sufficient time can be taken to allow complete separation to take place.

In a continuous production process, however, conditions are usually far from ideal and particles can be misplaced to the wrong product for a variety of reasons. The dominant effect is that of the density distribution of the feed. Very dense or very light particles will settle through the medium and report to the appropriate product quickly, but particles of density close to that of the medium will move more slowly and may not reach the right product in the time available for the separation. In the limit, particles of density the same as, or very close to, that of the medium will follow the medium and divide in much the same proportion.

Other factors also play a role in determining the efficiency of separation. Fine particles generally separate less efficiently than coarse particles because of their slower settling rates. The properties of the medium, the design and condition of the separating vessel, and the feed conditions, particularly feed rate, will all influence the separation.

(1) Partition Curve. The efficiency of separation can be represented by the slope of a partition or Tromp curve, first introduced by Tromp (1937). It describes the separating efficiency for the separator whatever the quality of the feed and can be used for estimation of performance and comparison between separators.

The partition curve relates the partition coefficient or partition value, that is, the percentage of the feed material of a particular specific gravity, which reports to either the sinks product (generally used for minerals) or the floats product (generally used for coal), to specific gravity (Fig. 5.15). It is exactly analogous to the classification efficiency curve (Chapter 3), in which the partition coefficient is plotted against particle size rather than specific gravity.

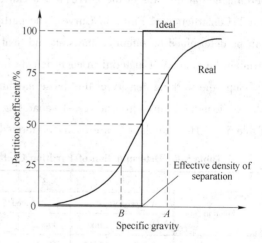

Fig. 5.15 Partition or tromp curve

The ideal partition curve reflects a perfect separation, in which all particles having a density higher than the separating density report to sinks and those lighter report to floats. There is no misplaced material.

The partition curve for a real separation shows that efficiency is highest for particles of density far from the operating density and decreases for particles approaching the operating density.

The area between the ideal and real curves is called the "error area" and is a measure of the degree of misplacement of particles to the wrong product.

Many partition curves give a reasonable straight-line relationship between the distribution of 25% and 75%, and the slope of the line between these distributions is used to show the efficiency of the process.

The probable error of separation or the Ecart probable (E_p) is defined as half the difference between the density where 75% is recovered to sinks and that at which 25% is recovered to sinks, that is, from Fig. 5.15:

$$E_\rho = \frac{A - B}{2} \tag{5.3}$$

The density at which 50% of the particles report to sinks is shown as the effective density of separation, which may not be exactly the same as the medium density, particularly for centrifugal separators, in which the separating density is generally higher than the medium density. This density of separation is referred to as the RD_{50} or ρ_{50} where the 50 refers to 50% chance of reporting to sinks (or floats).

The lower the E_ρ, the nearer to vertical is the line between 25% and 75% and the more efficient is the separation. An ideal separation has a vertical line with an $E_\rho = 0$, whereas in practice the E_ρ usually lies in the range 0.01~0.10.

The E_ρ is not commonly used as a method of assessing the efficiency of separation in units such as tables and spirals due to the many operating variables (wash water, table slope, speed, etc.) which can affect the separation efficiency. It is, however, ideally suited to the relatively simple and reproducible DMS process. However, care should be taken in its application, as it does not reflect performance at the tails of the curve, which can be important.

(2) Construction of Partition Curves. The partition curve for an operating dense medium vessel can be determined by sampling the sink and float products and performing heavy liquid tests to determine the amount of material in each density fraction. The range of liquid densities applied must envelope the working density of the dense medium unit. The results of heavy liquid tests on samples of floats and sinks from a vessel separating coal (floats) from shale (sinks) are shown in Table 5.4. The calculations are easily performed in a spreadsheet.

Table 5.4 Determination of Partition Coefficient for Vessel Separating Coal from Shale

Specific Gravity Fraction	1	2	3	4	5	6	7
	Analysis(wt)/%		of feed/%		Reconstituted Feed/%	Nominal Specific Gravity	Partition Coefficient
	Floats	Sinks	Floats	Sinks			
<1.30	83.34	18.15	68.84	3.16	71.98	—	4.39
1.30~1.40	10.50	10.82	8.67	1.88	10.56	1.35	17.84
1.40~1.50	3.35	9.64	2.77	1.68	4.45	1.45	37.74
1.50~1.60	1.79	13.33	1.48	2.32	3.80	1.55	61.07
1.60~1.70	0.30	8.37	0.25	1.46	1.71	1.65	85.46
1.70~1.80	0.16	5.85	0.13	1.02	1.15	1.75	88.51
1.80~1.90	0.07	5.05	0.06	0.88	0.94	1.85	93.83
1.90~2.00	0.07	4.34	0.06	0.76	0.81	1.95	92.89
>2.00	0.42	24.45	0.35	4.25	4.60	—	92.46
Total	100.00	100.00	82.60	17.40	100.00		

Columns 1 and 2 are the results of laboratory tests on the float and sink products, and columns 3 and 4 relate these results to the total distribution of the feed material to floats and sinks, which, in this example, is determined by directly weighing the two products over a period of time. The result of such determinations showed that 82.60% of the feed reported to the floats product (and 17.40% reported to the sinks) (see the Total row). Thus, for example, the first number in column 3 is (83.34/100)×82.60 (=63.84) and in column 4 it is (18.15/100)×17.40 (=3.16). The weight fraction in columns 3 and 4 can be added together to produce the reconstituted feed weight distribution in each density fraction (column 5). Column 6 gives the nominal (average) specific gravity of each density range, for example material in the density range 1.30~1.40 is assumed to have a specific gravity lying midway between these densities, that is, 1.35. Since the −1.30 specific gravity fraction and the +2.00 specific gravity fraction have no bound, no nominal density is given.

The partition coefficient (column 7) is the percentage of feed material of a certain nominal specific gravity which reports to sinks, that is:

$$\frac{\text{column 4}}{\text{column 5}} \times 100\%$$

The partition curve can then be constructed by plotting the partition coefficient against the nominal specific gravity, from which the separation density and probable error of separation of the vessel can be determined. The plot is shown in Fig. 5.16. Reading from the plot, the RD_{50} is ca. 1.52 and the E_p ca. 0.12 ((1.61−1.37)/2).

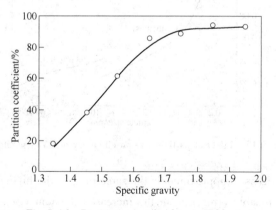

Fig. 5.16 Partition curve for data in Table 5.4

The partition curve can also be determined by applying the mass balancing procedure, provided the density distributions of feed, as well as sinks and floats, are available. Often the feed is difficult to sample, and thus resort is made to direct measurement of sinks and floats flowrates. It must be understood, however, that the mass balancing approach is the better way to perform the calculations as redundant data are available to execute data reconciliation. The mass balance/data reconciliation method is illustrated in determination of the partition curve for a hydrocyclone in Chapter 3.

An alternative, rapid, method of determining the partition curve of a separator is to use density tracers. Specially developed color-coded plastic tracers of known density can be fed to the process, the partitioned products being collected and hand sorted by density (color). It is then a simple matter to construct the partition curve directly by noting the proportion of each density of tracer reporting to either the sink or float product. Application of tracer methods has shown that considerable uncertainties can exist in experimentally determined Tromp curves unless an adequate number of tracers is used, and Napier Munn (1985) presents graphs that facilitate the selection of sample size and the calculation of confidence limits. A system in operation in a US coal preparation plant uses sensitive metal detectors that automatically spot and count the number of different types of tracers passing through a stream (Chironis, 1987).

Partition curves can be used to predict the products that would be obtained if the feed or separation density were changed. The curves are specific to the vessel for which they were established and are not affected by the type of material fed to it, provided:

1) The feed size range is the same-efficiency generally decreases with decrease in size; Fig. 5.17 shows typical efficiencies of gravitational separators or baths (drum, cone, etc.) and centrifugal separators (DMC, Dyna Whirlpool, etc.) versus particle size. It can be seen that, in general, below about 10mm, centrifugal separators are better than gravitational separators.

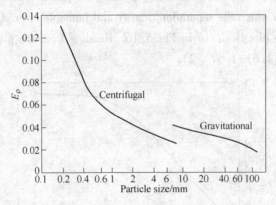

Fig. 5.17 Effect of particle size on efficiency of DM separators

2) The separating density is in approximately the same range—the higher the effective separating density the greater the probable error, due to the increased medium viscosity. It has been shown that the E_p is directly proportional to the separating density, all other factors being the same (Gottfried, 1978).

3) The feed rate is the same.

The partition curve for a vessel can be used to determine the amount of misplaced material that will report to the products for any particular feed material. For example, the distribution of the products from the tin ore, which was evaluated by heavy liquid tests (Table 5.1), can be determined for treatment in an operating separator. Fig. 5.18 shows a partition curve for a separator having an E_p of 0.07.

The curve can be shifted slightly along the abscissa until the effective density of separation corresponds to the laboratory evaluated separating density of 2.75. The distribution of material to sinks and floats can now be evaluated: for example, at a nominal specific gravity of 2.725, 44.0%

of the material reports to the sinks and 56.0% to the floats.

The performance is evaluated in Table 5.5. Columns 1,2,and 3 show the results of the heavy liquid tests, which were tabulated in Table 5.1. Columns 4 and 5 are the partition values to sinks and floats, respectively, obtained from the partition curve. Column 6 = column 1 ×column 4, and column 9 = column 1×column 5. The assay of each fraction is assumed to be the same, whether or not the material reports to sinks or floats (columns 2,7, and 10). Columns 8 and 11 are then calculated as the amount of tin reporting to sinks and floats in each fraction (columns 6×7 and 9×10) as a percentage of the total tin in the feed (sum of columns 1×2, i. e. ,1.12).

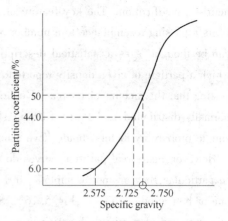

Fig. 5.18 Partition curve for $E_p = 0.07$

Table 5.5 Separation of Tin Ore Evaluation

Specific Gravity Fraction	Nominal S.G.	Feed			Partition Value/%		Predicted Sinks			Predicted Floats		
		1	2	3	4	5	6	7	8	9	10	11
		w/%	w(Sn)/%	Sn Dist/%	Sinks	Floats	w/%	w(Sn)/%	Sn Dist/%	w/%	w(Sn)/%	Sn Dist/%
<2.55	—	1.57	0.003	0.004	0.0	100.0	0.00	0.003	0.00	1.57	0.003	0.04
2.55~2.60	2.575	9.22	0.04	0.33	6.0	94.0	0.55	0.04	0.02	8.67	0.04	0.31
2.60~2.65	2.625	26.11	0.04	0.93	13.5	86.5	3.52	0.04	0.13	22.59	0.04	0.80
2.65~2.70	2.675	19.67	0.04	0.70	27.0	73.0	5.31	0.04	0.19	14.35	0.04	0.51
2.70~2.75	2.725	11.91	0.17	1.81	44.0	56.0	5.24	0.17	0.80	6.67	0.17	1.01
2.75~2.80	2.775	10.92	0.34	3.32	63.0	37.0	6.88	0.34	2.09	4.04	0.34	1.23
2.80~2.85	2.825	7.87	0.37	2.60	79.5	20.5	6.26	0.37	2.07	1.61	0.37	0.53
2.85~2.90	2.875	2.55	1.30	2.96	90.5	9.5	2.32	1.30	2.68	0.24	1.30	0.28
>2.90	—	10.18	9.60	87.34	100.00	0.00	10.18	9.60	87.31	0.00	9.60	0.00
Total		100.00	1.12	100.00			40.26	2.65	95.29	59.74	0.09	4.71

The total distribution of the feed to sinks is the sum of all the fractions in column 6, that is, 40.26%, while the recovery of tin into the sinks is the sum of the fractions in column 8, that is, 95.29%. This compares with a distribution of 31.52% and a recovery of 96.19% of tin in the ideal separation. In terms of upgrading, the grade of tin in the sinks is now 2.65% (solving Eq. (5.1), i. e. ,96.29 ×1.12/40.26) compared to the ideal of 3.42% Sn.

This method of evaluating the performance of a separator on a particular feed is tedious and is ideal for a spreadsheet, providing that the partition values for each density fraction are known. These can be represented by a suitable mathematical function. There is a large literature on the selection and application of such functions. Some are arbitrary, and others have some theoretical or heuristic justification. The key feature of the partition curve is its S-shaped character. In this it bears a passing resemblance to a number of probability distribution functions, and indeed the curve can be thought of as a statistical description of the DMS process, describing the probability with which a particle of given density reports to the sink product. Tromp himself recognized this in suggesting that the amount of misplaced material relative to a suitably transformed density scale was normally distributed, and Jowett (1986) showed that a partition curve for a process controlled by simple probability factors should have a normal distribution form.

However, many real partition curves do not behave ideally like the one illustrated in Fig. 5.15. In particular, they are not asymptotic to 0 and 100%, but exhibit evidence of short-circuit flow to one or both products (e.g., Fig. 5.16). Stratford and Napier-Munn (1986) identified four attributes required of a suitable function to represent the partition curve:

1) It should have natural asymptotes, preferably described by separate parameters.

2) It should be capable of exhibiting asymmetry about the separating density; That is, the differentiated form of the function should be capable of describing skewed distributions.

3) It should be mathematically continuous.

4) Its parameters should be capable of estimation by accessible methods.

A two-parameter function asymptotic to 0 and 100% is the Rosin-Rammler function, originally developed to describe size distributions (Tarjan, 1974):

$$P_i = 100 - 100\exp\left[-\left(\frac{\rho_i}{a}\right)^m\right] \tag{5.4}$$

In this form, P_i is the partition number (feed reporting to sinks, %), ρ_i the mean density of density fraction i, and a and m the parameters of the function; m describes the steepness of the curve (high values of m indicating more efficient separations). Partition curve functions are normally expressed in terms of the normalized density, ρ/ρ_{50}, where ρ_{50} is the separating density (RD_{50}). The normalized curve is generally independent of cut-point and medium density, but is dependent on particle size. Inserting this normalized density into Eq. (5.4), and noting that $P=50$ for $\rho=\rho_{50}(\rho/\rho_{50}=1)$, gives:

$$P_i = 100 - 100\exp\left[-\ln2\left(\frac{\rho_i}{\rho_{50}}\right)^m\right] \tag{5.5}$$

One of the advantages of Eq. (5.5) is that it can be linearized so that simple linear regression can be used to estimate m and ρ_{50} from experimental data:

$$\ln\left[\frac{\ln\left(\frac{100}{100-P_i}\right)}{\ln2}\right] = m\ln\rho_i - m\ln\rho_{50} \tag{5.6}$$

(This approach is less important today with any number of curve-fitting routines available (and Excel Solver), the same point also made in Chapter 3 when curve-fitting cyclone partition curves.)

Gottfried (1978) proposed a related function, the Weibull function, with additional parameters to account for the fact that the curves do not always reach the 0 and 100% asymptotes due to short-circuit flow:

$$P_i = 100 - 100\left\{f_0 + c \exp\left[-\frac{\left(\frac{\rho_i}{\rho_{50}} - x_0\right)^a}{b}\right]\right\} \tag{5.7}$$

The six parameters of the function ($c, f_0, \rho_{50}, x_0, a$, and b) are not independent, so by the argument of Eq. (5.5), x_0 can be expressed as:

$$x_0 = 1 - \left[b \ln\left(\frac{c}{0.5 - f_0}\right)\right]^{\frac{1}{a}} \tag{5.8}$$

In this version of the function, representing percentage of feed to sinks, f_0 is the proportion of high-density material misplaced to floats, and $1-(c+f_0)$ is the proportion of low-density material misplaced to sinks, so that $c+f_0 \leq 1$. The curve therefore varies from a minimum of $100[1-(c+f_0)]$ to a maximum of $100(1-f_0)$.

The parameters of Eq. (5.8) have to be determined by nonlinear estimation. First approximations of c, f_0, and ρ_{50} can be obtained from the curve itself.

King and Juckes (1988) used Whiten's classification function (Lynch, 1977) with two additional parameters to describe the short-circuit flows or by-pass:

$$P_i = \beta + (1 - \alpha - \beta)\left[\frac{\exp\left(\frac{b\rho_i}{\rho_{50}}\right) - 1}{\exp\left(\frac{b\rho_i}{\rho_{50}}\right) + \exp(b) - 2}\right] \tag{5.9}$$

here, for P_i is the proportion to underflow, α the fraction of feed which short-circuits to overflow, and β the fraction of feed which short-circuits to underflow; b is an efficiency parameter, with high values of b indicating high efficiency. Again, the function is nonlinear in the parameters.

The E_ρ can be predicted from these functions by substitution for ρ_{75} and ρ_{25}. Scott and Napier-Munn (1992) showed that for efficient separations (low E_ρ) without shortcircuiting, the partition curve could be approximated by:

$$P_i = \frac{1}{1 + \exp\left(\frac{\ln 3(\rho_{50} - \rho_i)}{E_\rho}\right)} \tag{5.10}$$

(3) Organic Efficiency. This term is often used to express the efficiency of coal preparation plants. It is defined as the ratio (normally expressed as a percentage) between the actual yield of a desired product and the theoretical possible yield at the same ash content.

For instance, if the coal, whose washability data are plotted in Fig. 5.13, produced an operating yield of 51% at an ash content of 12%, then, since the theoretical yield at this ash content is 55%, the organic efficiency is equal to 51/55 or 92.7%.

Organic efficiency cannot be used to compare the efficiencies of different plants, as it is a dependent criterion, and is much influenced by the washability of the coal. It is possible, for example, to obtain a high organic efficiency on a coal containing little near-density material, even when the separating efficiency, as measured by partition data, is quite poor.

REFERENCES

Baguley, P. J., Napier-Munn, T. J., 1996. Mathematical model of the dense medium drum. Trans. Inst. Min. Metall. Sec. C. 105-106, C1-C8.

Banerjee, P. K., et al., 2003. A plant comparison of the vorsyl separator and dense medium cyclone in the treatment of Indian coals. Int. J. Miner. Process. 69 (1-4), 101-114.

Bekker, E., 2014. DMC basics: a holistic view. Proceedings of the 11th DMS Powders Conference, Mount Grace, South Africa.

Burton, M. W. A., et al., 1991. The economic impact of modern dense medium systems. Miner. Eng. 4 (3-4), 225-243.

Cebeci, Y., Ulusoy, U., 2013. An optimization study of yield for a coal washing plant from Zonguldak region. Fuel Process. Technol. 115, 110-114.

Chironis, N. P., 1987. On-line coal-tracing system improves cleaning efficiencies. Coal Age. 92 (Mar.), 44-47.

Collins, B., et al., 1974. The production, properties and selection of ferrosilicon powders for heavy medium separation. J. S. Afr. Inst. Min. Metall. 75 (5), 103-119.

Dardis, K. A., 1987. The design and operation of heavy medium recovery circuits for improved medium recovery. Proceedings of Dense Medium Operator's Conference. AusIMM, Brisbane, Australia, 157-184.

Davis, J. J., 1992. Cleaning coarse and small coal—dense medium processes. In: Swanson, A. R., Partridge, A. C. (Eds.), Advanced Coal Preparation Monograph Series, vol. 3. Australian Coal Preparation Society, Broadmeadow, NSW, Australia, part 8.

Domenico, J. A., et al., 1994. Magstreamsas a heavy liquid separation alternative for mineral sands exploration. SME Annual Meeting. SME, Albuquerque, NM, USA, Preprint 94-262.

Dunglison, M. E., et al., 2000. The rheology of ferrosilicon dense medium suspensions. Miner. Process. Extr. Metall. Rev. 20 (1), 183-196.

Ferrara, G., et al., 1994. Tri-Flo: a multistage high-sharpness DMS process with new applications. Miner. Metall. Process. 11 (2), 63-73.

Gottfried, B. S., 1978. A generalisation of distribution data for characterizing the performance of float-sink coal cleaning devices. Int. J. Miner. Process. 5 (1), 1-20.

Hacioglu, E., Turner, J. F., 1985. A study of the Dyna Whirlpool, Proceedings of 15th International Mineral Processing Congress, vol. 1. Cannes, France, 244-257.

Jowett, A., 1986. An appraisal of partition curves for coal-cleaning processes. Int. J. Miner. Process. 16 (1-2), 75-95.

King, R. P., 2012. In: Schneider, C. L., King, E. A. (Eds.), Modeling and Simulation of Mineral Processing Systems, second ed. SME, Englewood, Co., USA.

King, R. P., Juckes, A. H., 1988. Performance of a dense medium cyclone when beneficiating fine coal. Coal Preparation. 5 (3-4), 185-210.

Kitsikopoulos, H., et al., 1991. Industrial operation of the first twodensity three-stage dense medium separator processing chromite ores. Proceedings of the 17th International Mineral Processing Congress (IMPC), vol. 3. Dresden, Freiberg, Germany, 55-66.

Kuang, Sh., et al., 2014. CFD modeling and analysis of the multiphase flow and performance of dense medium cyclones. Miner. Eng. 62,43-54.

Lane, D. E., 1987. Point of Ayr Colliery. Mining Mag. 157 (Sept.),226-237.

Leonard, J. W. (Ed.),1991. Coal Preparation. SME, Littleton, Co., USA. Lin, D., et al., 1995. Batch magnetohydrostatic separations in a modified Frantz Separator. Miner. Eng. 8 (3),283-292.

Luttrell, G. H., 2014. Innovations in coal processing. In: Anderson, C. G., et al., (Eds.), Mineral Processing and Extractive Metallurgy: 100 Years of Innovation. SME, Englewood, Co., USA, 277-296.

Lynch, A. J., 1977. Mineral Crushing and Grinding Circuits: Their Simulation, Optimisation, Design and Control. Elsevier Scientific Publishing Company, Amesterdam, the Netherlands.

Majumder, A. K., et al., 2009. Applicability of a dense-medium cyclone and Vorsyl Separator for upgrading non-coking coal fines for use as a blast furnace injection fuel. Int. J. Coal Prep. Utils. 29 (1),23-33.

Mills, C., 1980. Process design, scale-up and plant design for gravity concentration. In: Mular, A. L., Bhappu, R. B. (Eds.), Mineral Processing Plant Design, second ed. AIMME, New York, NY, USA, 404-426 (Chapter 18).

Munro, P. D., et al., 1982. The design, construction and commissioning of a heavy medium plant of silver-lead-zinc ore treatment-Mount Isa Mines Limited. Proceedings of the 14th International Mineral Processing Congress, Toronto, ON, Canada, VI-6. 1VI-6. 20.

Myburgh, H. A., 2002. The influence of the quality of ferrosilicon on the rheology of dense medium and the ability to reach higher densities. Proceedings of Iron Ore 2002. AusIMM, Perth, WA, USA, pp. 313-317.

Napier-Munn, T. J., 1985. Use of density tracers for determination of the Tromp curve for gravity separation processes. Trans. Inst. Min. Metall. 94 (Mar.), C45-C51.

Napier-Munn, T. J., 1990. The effect of dense medium viscosity on separation efficiency. Coal Preparation. 8 (3-4),145-165.

Napier-Munn, T. J., Scott, I. A., 1990. The effect of demagnetisation and ore contamination on the viscosity of the medium in a dense medium cyclone plant. Miner. Eng. 3 (6),607-613.

Napier-Munn, T. J., et al., 1995. Some causes of medium loss in dense medium plants. Miner. Eng. 8 (6), 659-678.

Napier-Munn, T. J., et al., 2009. Advances in dense-medium cyclone plant design. Proceedings of the 10th Mill Operation' Conference. AusIMM, Adelaide, SA, Australia, 53-61.

Napier-Munn, T. J., et al., 2014. Innovations in dense medium separation technology. In: Anderson, C. G., et al., (Eds.), Mineral Processing and Extractive Metallurgy: 100 Years of Innovation. SME, Englewood, Co., USA, 265-276.

Osborne, D., 2010. Value of R&D in coal preparation development. In: Honaker, R. Q. (Ed.), Proceedings of the 16th International Coal Preparation Congress. SME, Lexington, KY, USA, 845-858.

Parsonage, P., 1980. Factors that influence performance of pilot-plant paramagnetic liquid separator for dense particle fractionation. Trans. Inst. Min. Metall. Sec. C. 89 (Dec.), C166-C173.

Pascoe, R. D., Hou, Y. Y., 1999. Investigation of the importance of particle shape and surface wettability on the separation of plastics in a LARCODEMS separator. Miner. Eng. 12 (4),423-431.

Plewinsky, B., Kamps, R., 1984. Sodium metatungstate, a new medium for binary and ternary density gradient centrifugation. Die Makromolekulare Chemie. 185 (7),1429-1439.

Reeves, T. J., 1990. On-line viscosity measurement under industrial conditions. Coal Prep. 8 (3-4),135-144.

Richard, G. M., et al., 2011. Optimization of the recovery of plastics for recycling by density media separation cyclones. Resour. Conserv. Recy. 55(4),472-482.

Ruff, H. J., 1984. New developments in dynamic dense-medium systems. Mine Quarry. 13 (Dec.),24-28.

Rylatt, M. G., Popplewell, G. M., 1999. Diamond processing at Ekati in Canada. Mining Eng. 51 (2), 19-25.

Schena, G. D., et al., 1990. Pre-concentration by dense-medium separation—an economic evaluation. Trans. Inst. Min. Metall. Sec. C. 99-100 (Jan. -Apr.), C21-C31.

Scott, I. A., Napier-Munn, T. J., 1992. Dense medium cyclone model based on the pivot phenomenon. Trans. Inst. Min. Metall. 101 (Jan. Apr.), C61-C76.

Shah, C. L., 1987. A new centrifugal dense medium separator for treating 250t/h of coal sized up to 100mm. In: Wood, P. (Ed.), Proceedings of the Third International Conference on Hydrocyclones. BHRA the Fluid Engineering Center, Oxford, England, 91-100.

Stewart, K. J., Guerney, P. J., 1998. Detection and prevention of ferrosilicon corrosion in dense medium plants. Proceedings of the Sixth Mill Operators' Conference. AusIMM, Madang, Papua New Guinea, 177-183.

Stratford, K. J., Napier-Munn, T. J., 1986. Functions for the mathematical representation of the partition curve for dense medium cyclones. Proceedings of the 19th Application of Computers & Operations Research in the Mineral Industry (APCOM). SME, Pittsburgh, PA, USA, 719-728.

Symonds, D. F., Malbon, S., 2002. Sizing and selection of heavy media equipment: design and layout. In: Mular, Halbe, Barratt (Eds.), Mineral Processing Plant Design, Practice and Control, vol. 1. SME, Littleton, Co., USA, 1011-1032.

Tarjan, G., 1974. Application of distribution functions to partition curves. Int. J. Miner. Process. 1 (3), 261-264.

Tromp, K. F., 1937. New methods of computing the washability of coals. Gluckauf. 73, 125-131.

Williams, R. A., Kelsall, G. H., 1992. Degradation of ferrosilicon media in dense medium separation circuits. Miner. Eng. 5 (1), 57-77.

Wills, B. A., Lewis, P. A., 1980. Applications of the Dyna Whirlpool in the minerals industry. Mining Mag. 143 (3), 255-257.

Chapter 6   Magnetic and Electrical Separation

6.1   INTRODUCTION

Magnetic and electrical separators are being considered in the same chapter, as there is often an overlap in the application of the two processes. A classic example of this is the processing of heavy mineral sands in which both magnetic and electrostatic separation are crucial to achieve separation.

6.2   MAGNETISM IN MINERALS

Magnetic separators exploit the difference in magnetic properties between the minerals in a deposit and are used to concentrate a valuable mineral that is magnetic (e.g., magnetite from quartz), to remove magnetic contaminants, or to separate mixtures of magnetic and nonmagnetic valuable minerals. An example of the latter is the tin-bearing mineral cassiterite, which is often associated with traces of the valuable minerals magnetite or wolframite, which can be removed by magnetic separators.

This text will only briefly introduce the concepts associated with magnetism in mineral separation. For those interested in further details there are a number of other sources (Jiles, 1990; Oberteuffer, 1974; Svoboda, 1987; Svoboda and Fujita, 2003).

All materials are affected in some way when placed in a magnetic field, although with many substances the effect is too slight to be easily detected. For the purposes of mineral processing, materials may be classified into two broad groups, according to whether they are attracted or repelled by a magnet:

(1) Diamagnetic materials are repelled along the lines of magnetic force to a point where the field intensity is smaller. The forces involved here are very small and diamagnetic substances are often referred to as "non-magnetic", although this is not strictly correct. Diamagnetic minerals will report to the nonmagnetic product ("non-mags") of a magnetic separator as they do not experience a magnetic attractive force.

(2) Paramagnetic materials are attracted along the lines of magnetic force to points of greater field intensity. Paramagnetic materials report to the "magnetic" product ("mags") of a magnetic separator due to attractive magnetic forces. Examples of paramagnetic minerals which are separated in commercial magnetic separators are ilmenite ($FeTiO_3$), rutile (TiO_2), wol-framite ($(Fe,Mn)WO_4$), monazite ($(Ce,La,Nd,Th)PO_4$), xenotime (YPO_4), siderite ($FeCO_3$),

chromite ($FeCr_2O_4$), and manganese minerals.

Paramagnetism in a material originates due to the presence of unpaired electrons which create magnetic dipoles. When these magnetic dipoles are aligned by an externally applied magnetic field, the resultant magnetic moment causes the material to become magnetized and experience a magnetic force along the lines of the applied magnetic field. Certain elements have electron configurations with many unpaired electrons, but the magnetic response of a given mineral depends on the structure of the mineral as well as its constituent atoms. For example, pyrite (FeS_2) is very slightly paramagnetic, but the chemically similar pyrrhotite ($Fe_{1-x}S$) in the monoclinic structural form is actually strongly magnetic, referred to as ferromagnetic.

Ferromagnetism can be regarded as a special case of paramagnetism in which the magnetic dipoles of a material undergo exchange coupling so that they can more rapidly align themselves with an applied magnetic field. Examples of diamagnetic, paramagnetic, and ferromagnetic behavior are shown in Fig. 6.1 and Fig. 6.2 represented as magnetization (density of magneticdipoles) versus applied magnetic field strength. The slope of these curves represents the dimensionless magnetic susceptibility of the material. Fig. 6.1 shows the para-magnetic susceptibility (shown by a positive linear slope) of chromite and the diamagnetic susceptibility (negative linear slope) of quartz, while Fig. 6.2 shows the ferromagnetic trend of magnetite. Thanks to exchange coupling, ferromagnetic materials will have very high initial susceptibility to magnetic forces until all of the exchange coupled magnetic moments have aligned with the applied magnetic force. This results in a rapidly decreasing value of susceptibility with increased applied magnetic field (Fig. 6.2, points 1,3). Once this alignment has occurred the material is said to have reached its saturation magnetization (a characteristic value of the material shown in Fig. 6.2 as a plateau in magnetization), and any further increase in applied magnetic field will not be accompanied by a further increase in magnetization.

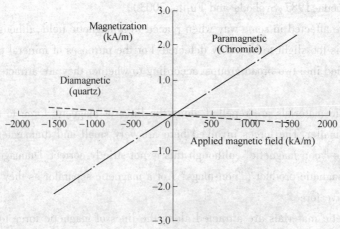

Fig. 6.1 Magnetization versus applied magnetic field strength for idealized paramagnetic and diamagnetic minerals

Compared to paramagnetic materials, which need high-intensity (high magnetic field) magnetic separators to report to the magnetic product, ferromagnetic materials are recovered in low-intensity

6.2 MAGNETISM IN MINERALS

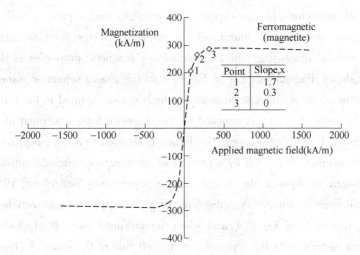

Fig. 6.2 Magnetization versus applied magnetic field strength for a ferromagnetic mineral

magnetic separators.

Measuring Magnetic Properties: The magnetic properties of a material may be measured directly via a vibrating sample magnetometer (used to obtain the data for Fig. 6.1 and Fig. 6.2) which is specifically designed to capture the variation of the magnetic properties of a material as a function of applied magnetic field strength. Empirical data for paramagnetic mineral samples often includes both paramagnetic and ferromagnetic characteristics due to the presence of impurities in the sample, as mineral grains, or in the mineral's crystal structure. Similarly, measurements of a predominantly diamagnetic sample may show signs of a paramagnetic impurity. In Fig. 6.3, the results are shown for the samples from Fig. 6.1 prior to data processing to isolate the paramagnetic or diamagnetic trend.

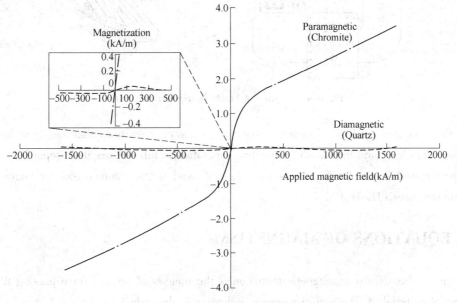

Fig. 6.3 Magnetization versus applied magnetic field strength for natural samples

Vibrating sample magnetometers are expensive instruments and require specialized personnel to operate. As such, they are typically found only in research settings, such as universities. A more practical tool in mineral processing labs for determining magnetic properties is the Frantz Isodynamic Separator, shown diagrammatically in Fig. 6. 4. In the Frantz separator, mineral particles are fed down a vibrating chute inclined at an angle θ_2 which is also inclined in the transverse direction at an angle θ_1. The force of gravity is opposed by the magnetic force generated by the electromagnetic coil through which the chute passes. The Frantz is referred to as an isodynamic separator due to the fact that the magnetic force felt by a particle of constant magnetic susceptibility and orientation remains constant throughout the length of the separator (McAndrew, 1957). When the attractive magnetic force is stronger than the force of gravity on a mineral particle it will report to the chute on the right side of Fig. 6. 4, and when the magnetic force is insufficient to overcome gravity the mineral particle exits the separator on the left side of the chute. As the current through the electromagnetic coil and both θ_1 and θ_2 may be varied across a wide range, this separator is able to separate minerals of varying magnetic properties. It may even be used to concentrate diamagnetic minerals, in which case the side slope is moved past horizontal such that the chute exit on the right of Fig. 6. 4 becomes the down slope exit for particles where the diamagnetic force (repulsive magnetic force) is insufficient to overcome the force of gravity. The Frantz may also be used to determine the magnetic susceptibility of a given mineral, provided other materials of known susceptibility are available for proper calibration (McAndrew, 1957). Normally operated dry, it can be modified to operate wet (Todd and Finch, 1984).

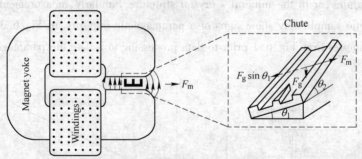

Fig. 6. 4 Diagram of Frantz Isodynamic Separator

The Franz is also used to characterize materials, separating fractions that can then be identified. These data can form the basis of predicting separation in full size magnetic separators. For ferromagnetic materials, the Davis tube is more suitable and is the common tool for characterizing magnetic iron ores (Davis, 1921).

6. 3   EQUATIONS OF MAGNETISM

The magnetic flux density or magnetic induction is the number of lines of force passing through a unit area of material, B. The unit of magnetic induction is the tesla (T).

The magnetizing force, which induces the lines of force through a material, is called the field intensity, H (or H-field), and by convention has the units ampere per meter (A/m) (Bennett et al., 1978).

The intensity of magnetization or the magnetization (M, A/m) of a material relates to the magnetization induced in the material and can also be thought of as the volumetric density of induced magnetic dipoles in the material. The magnetic induction, B, field intensity, H, and magnetization, M, are related by the equation:

$$B = \mu_0(H + M) \tag{6.1}$$

where μ_0 is the permeability of free space and has the value of $4\pi \times 10^{-7} \text{N/A}^2$. In a vacuum, $M = 0$, and M is extremely low in air and water, such that for mineral processing purposes Eq. (6.1) may be simplified to:

$$B = \mu_0 H \tag{6.2}$$

so that the value of the field intensity, H, is directly proportional to the value of induced flux density, B (or B-field), and the term "magnetic field intensity" is then often loosely used for both the H-field and the B-field. However, when dealing with the magnetic field inside materials, particularly ferromagnetic materials that concentrate the lines of force, the value of the induced flux density will be much higher than the field intensity. This relationship is used in high-gradient magnetic separation (discussed further in Section 6.4.1). For clarity it must be specified which field is being referred to.

Magnetic susceptibility (χ) is the ratio of the intensity of magnetization produced in the material over the applied magnetic field that produces the magnetization:

$$\chi = \frac{M}{H} \tag{6.3}$$

Combining Eqs. (6.1) and (6.3) we get:

$$B = \mu_0 H(1 + \chi) \tag{6.4}$$

If we then define the dimensionless relative permeability, μ, as:

$$\mu = 1 + \chi \tag{6.5}$$

We can combine Eqs. (6.4) and (6.5) to yield:

$$B = \mu\mu_0 H \tag{6.6}$$

For paramagnetic materials, χ is a small positive constant, and for diamagnetic materials it is a much smaller negative constant. As examples, from Fig. 6.1 the slope representing the magnetic susceptibility of the material, χ, is about 0.001 for chromite and -0.0001 for quartz.

The magnetic susceptibility of a ferromagnetic material is dependent on the magnetic field, decreasing with field strength as the material becomes saturated. Fig. 6.2 shows a plot of M versus H for magnetite, showing that at an applied field of 80kA/m, or 0.1T, the magnetic susceptibility is about 1.7, and saturation occurs at an applied magnetic field strength of about 500kA/m or 0.63T. Many high-intensity magnetic separators use iron cores and frames to produce the desired magnetic flux concentrations and field strengths. Iron saturates magnetically at about 2~2.5T, and its non-linear ferromagnetic relationship between inducing field strength and magnetization intensity necessitates the use of very large currents in the energizing coils, sometimes up to hundreds of amperes.

The magnetic force felt by a mineral particle is dependent not only on the value of the field intensity, but also on the field gradient (the rate at which the field intensity increases across the particle toward the magnet surface). As paramagnetic minerals have higher (relative) magnetic permeabilities than the surrounding media, usually air or water, they concentrate the lines of force of an external magnetic field. The higher the magnetic susceptibility, the higher the induced field density in the particle and the greater is the attraction up the field gradient toward increasing field strength. Diamagnetic minerals have lower magnetic susceptibility than their surrounding medium and hence expel the lines of force of the external field. This causes their expulsion down the gradient of the field in the direction of the decreasing field strength.

The equation for the magnetic force on a particle in a magnetic separator depends on the magnetic susceptibility of the particle and fluid medium, the applied magnetic field and the magnetic field gradient. This equation, when considered in only the x-direction, may be expressed as (Oberteuffer, 1974):

$$F_x = V(\chi_p - \chi_m) H \frac{dB}{dx} \quad (6.7)$$

where F_x is the magnetic force on the particle (N), V the particle volume (m^3), χ_p the magnetic susceptibility of the particle, χ_m the magnetic susceptibility of the fluid medium, H the applied magnetic field strength (A/m), and dB/dx the magnetic field gradient ($T/m = N/(A \cdot m^2)$). The product of H and dB/dx is sometimes referred to as the "force factor".

Production of a high field gradient as well as high intensity is therefore an important aspect of separator design. To generate a given attractive force, there are an infinite number of combinations of field and gradient which will give the same effect. Another important factor is the particle size, as the magnetic force experienced by a particle must compete with various other forces such as hydrodynamic drag (in wet magnetic separations) and the force of gravity. In one example, considering only these two competing forces, Oberteuffer (1974) has shown that the range of particle size where the magnetic force predo-minates is from about 5μm to 1mm.

6.4 MAGNETIC SEPARATOR DESIGN

6.4.1 Magnetic Field Gradient

Certain elements of design are incorporated in all magnetic separators, whether they are wet or dry separators with low- or high-intensity magnetic fields. The prime requirement is the provision of a high-intensity field in which there is a steep field strength gradient. In a field of uniform magnetic flux, such as in Fig. 6.5(a), magnetic particles will orient, but will not move along the lines of flux. The most straightforward method for producing a converging field is by providing a V-shaped pole above a flat pole, as in Fig. 6.5(b). The tapering of the upper pole concentrates the magnetic flux into a very small area, giving high intensity. The lower flat pole has the same total magnetic flux distributed over a larger area. Thus, there is a steep field gradient across the gap by virtue of the different field intensity levels. Another method of producing a high field gradient is by using a pole which is constructed of alternating magnetic and nonmagnetic laminations (Fig. 6.6).

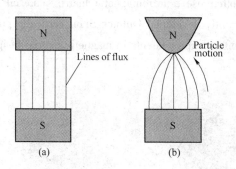
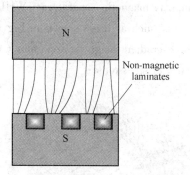

Fig. 6.5 (a) Field of uniform flux, and
(b) converging field generating force on particle

Fig. 6.6 Production of field gradient by laminated pole

The design of field gradient in magnetic separators may be divided into two types: open-gradient magnetic separators (OGMSs) and high-gradient magnetic separators (HGMSs). In an OGMS design, the magnetic gradient is created by the poles of the magnets themselves, and as a result the gradient is relatively weak (Kopp, 1991). These type of separators include free-fall separators and drum separators where particles passing by the separator are deflected into different streams based on their magnetic properties (Kopp, 1991). In HGMS, a ferromagnetic matrix element is introduced into the applied magnetic field to create many points of high field gradient with the intention of capturing the magnetic particles and allowing nonmagnetic particles to flow through the separator.

The introduction of particles with high magnetic susceptibility into a magnetic field will concentrate the lines of force so that they pass through the particles themselves (Fig. 6.7). Since the lines of force converge to the particles, a high field gradient is produced, which causes the particles themselves to behave as magnets, thus attracting each other. Flocculation, or agglomeration, of the particles can occur if they are small and highly susceptible and if the field is intense. This is important

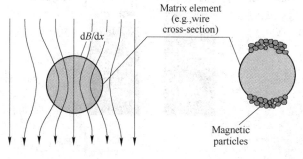

Fig. 6.7 Production of filed gradient by strongly magnetic matrix material, and consequent capture and buildup of magnetic particles

as these magnetic "flocs" can entrain gangue mineral particles as well as bridge the gaps between magnetic poles, reducing the efficiency of separation.

Much of the optimization of high-intensity separators is based on providing as many sites of high field gradient as possible to improve the magnetic particle carrying capacity of the separator. Wet high-intensity magnetic separators (WHIMSs) will often use a ferromagnetic matrix material to achieve this, such as those shown in Fig. 6.8. The recently developed Outotec SLon vertically pulsating high-gradient magnetic separator (VPHGMS) offers improvements in magnetic matrix design via the use of steel rods.

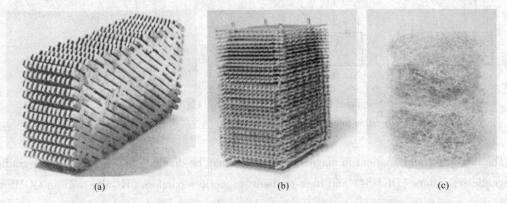

Fig. 6.8 Examples of matrix materials used in high-intensity separators:
(a) Section through Boxmag-Rapid grid assembly showing matrix of stainless steel bars,
(b) grid of expanded ferromagnetic stainless steel used for coarse particle sizes, and
(c) ferromagnetic stainless steel wool used for fine particle sizes (Courtesy Metso)

A visual comparison of the effects of different matrix materials on magnetic flux may be seen in Fig. 6.9. The design of the matrix can be further optimized by tapering the size and spacing of the

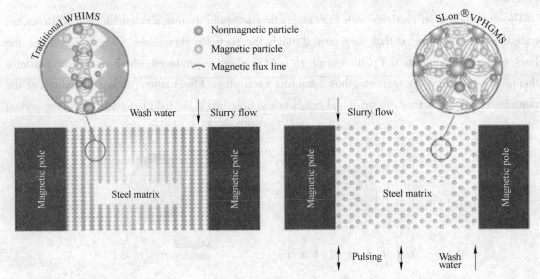

Fig. 6.9 Comparison of effects of magnetic matrix design on magnetic flux in traditional WHIMS and SLon VPHGMS (Courtesy Outotec)

rods throughout the matrix (Fig. 6.10) so that coarse magnetic particles are trapped first near the slurry inlet, with additional points of high field gradient introduced further along the direction of slurry flow to capture finer magnetic particles (Novotny,2014). Further details on WHIMS and VPHGMS may be found in Sections 6.5.2 and 6.5.3, respectively.

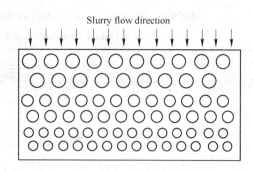

Fig. 6.10 Design of magnetic matrix to optimize capacity as well as maximize recovery of fine magnetic particles (Adapted from Novotny (2014))

6.4.2 Magnetic Field Intensity

Provision must be incorporated in the separator for regulating the intensity of the magnetic field in order to deal with various types of material. This is easily achieved in electromagnetic separators by varying the current, while with permanent magnets the interpole distance can be varied. In the case of laboratory separators, this can also be achieved by interchanging the permanent magnets for magnets of higher magnetic field intensity. A special class of magnetic separators, known as superconducting separators, may be used when very high field intensities are required. Additional information on superconducting separators may be found in Section 6.5.4.

It is important to note that increasing field intensity does not necessarily lead to an improved separation. Work by Svoboda (1994) with HGMS has shown that the magnetic field strength should be carefully selected according to the application, as higher field strengths may lead to increased capture of weakly magnetic gangue particles. The field should be sufficient to ensure that a particle which collides with a matrix element will remain fixed to that element; any further increase serves only to retain particles with weaker magnetic properties. Another negative impact of high field strength is that for particles exhibiting some degree of magnetic ordering, a relatively common situation in mineral processing, increased field strength actually serves to decrease the magnetic susceptibility (Svoboda,1994). Work by Shao et al. (1996) to measure the magnetic susceptibility of iron minerals at varying field strengths showed that from 0.4T to 0.9T the susceptibility of a hematite sample decreased by more than 50%. A similar result is seen in Fig. 6.2, where the susceptibility of magnetite decreases from 1.7 to 0.1 as the applied magnetic field strength is increased from 0.1 to 0.4T (80~320kA/m). While such a decrease in magnetic susceptibility is significant, it must be considered in the context of the increasing applied magnetic field strength, as both H and χ affect the force experienced by a mineral particle (Eq. (6.7)). The product of the two, χH, should be calculated to capture the effect of both the decreasing magnetic susceptibility and the increasing applied magnetic field strength. An example of such a calculation for the mineral from Fig. 6.2 is given in Table 6.1.

Table 6.1 Comparison of Product of Magnetic Susceptibility and Applied Magnetic Field Between Points Along the Curve in Fig. 6.2

Point	Applied magnetic field/kA · m^{-1}	χH/kA · m^{-1}
1	80	134
2	160	49
3	320	23

As magnetic field strength is increased, the magnetic field gradient in the separator will also change; this is not considered in the calculations in Table 6.1, although it will have a direct effect on the force experienced by the mineral particles. Excess applied field strength may actually decrease the field gradient in a given separator (Section 6.4.1) (Svoboda, 1994). Since the magnetic force on a particle is directly proportional to the magnetic susceptibility of the particle as well as the magnetic field gradient in the separator, as seen in Eq. (6.7), the net effect of increased field strength can actually be a decrease in the magnetic force experienced by the particle. It is therefore crucial that the appropriate magnetic field is applied for a given separation.

6.4.3 Material Transport in Magnetic Separators

Commercial magnetic separators are continuous-process machines, and separation is carried out on a moving stream of particles passing into and through the magnetic field. Close control of the speed of passage of the particles through the field is essential, which typically rules out free fall as a means of feeding. Belts or drums are very often used to transport the feed through the field.

As discussed in Section 6.4.1, flocculation of magnetic particles is a concern in magnetic separators, especially with dry separators processing fine material. If the ore can be fed through the field in a monolayer, this effect is much less serious, but, of course, the capacity of the machine is drastically reduced. Flocculation is often minimized by passing the material through consecutive magnetic fields, which are usually arranged with successive reversals of the polarity. This causes the particles to turn through 180°, each reversal tending to free the entrained gangue particles. The main disadvantage of this method is that flux tends to leak from pole to pole, reducing the effective field intensity.

Provision for collection of the magnetic and nonmagnetic fractions must be incorporated into the design of the separator. Rather than allow the magnetics to contact the pole-pieces, which then requires their detachment, most separators are designed so that the magnetics are attracted to the pole-pieces, but come into contact with some form of conveying device, which carries them out of the influence of the field, into a bin or a belt. Nonmagnetic disposal presents no problems; free fall from a conveyor into a bin is often used. Middlings are readily produced by using a more intense field after the removal of the highly magnetic fraction.

6.5 TYPES OF MAGNETIC SEPARATOR

Magnetic separators are generally classified into low-and high-intensity machines, but here we include high-gradient and superconducting devices.

6.5.1 Low-Intensity Magnetic Separators

Low-intensity separators are used to treat ferromagnetic materials and some highly paramagnetic minerals.

As shown in Fig. 6.2, minerals with ferromagnetic properties have high susceptibility at low applied field strengths and can therefore be concentrated in low intensity ($<\sim 0.3T$) magnetic separators. For low-intensity drum separators (Fig. 6.11) used in the iron ore industry, the standard field, for a separator with ferrite-based magnets, is 0.12T at a distance of 50mm from the drum surface (Novotny, 2014). Work by Murariu and Svoboda (2003) has also shown that such separators have maximum field strengths on the drum surface of less than 0.3T. The principal ferromagnetic mineral concentrated in mineral processing is magnetite (Fe_3O_4), although hematite (Fe_2O_3) and siderite ($FeCO_3$) can be roasted to produce magnetite and hence give good separation in low-intensity machines.

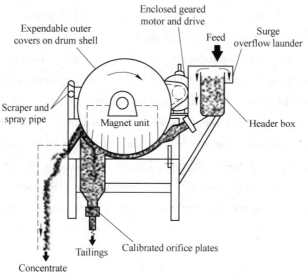

Fig. 6.11 Diagram of a typical drum separator

The removal of "tramp" iron from feed belts can also be regarded as a form of low-intensity magnetic separation. However, tramp iron removal is usually accomplished by means of a magnetic pulley at the end of an ore conveyor (Fig. 6.12) or by a guard magnet suspended over the conveyor belt (see Chapter 2). Tramp iron removal is important prior to crushing and in certain cases removal of the iron produced from grinding media wear can be important for downstream processing. A common example of the latter is processing gold ores, where the use of magnetic separation (typically a drum separator) in advance of centrifugal gravity concentration is used to remove tramp iron and prevent damage to the centrifugal separator, as well as avoiding contamination of the gravity concentrate with dense iron particles (Bird and Briggs, 2011).

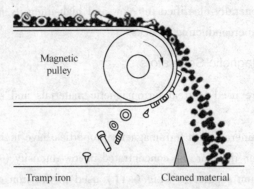

Fig. 6.12 Example of magnetic pulley used to remove tramp iron from an ore prior to further processing (Courtesy Eriez)

Dry low-intensity magnetic separation is confined mainly to the concentration of coarse sands which are strongly magnetic, a process known as "cobbing" and is often carried out using drum separators. For particles below 5mm, dry separation tends to be replaced by wet methods, which produce less dust loss and usually yield a cleaner product. Low-intensity wet separation is widely used for recycling (and cleaning) magnetic media in dense medium separation (DMS) processes (see Chapter 5) and for the processing of ferromagnetic sands.

The general design of drum separators is a rotating, hollow, nonmagnetic drum containing multiple stationary magnets of alternating polarity. The medium-intensity Permos separator uses many small magnet blocks, whose direction of magnetization changes in small steps. This is said to generate a very even magnetic field, requiring less magnetic material (Wasmuth and Unkelbach, 1991). The spatial arrangement of the magnets within a drum separator may be varied depending on the specific requirements of the application. This is illustrated by the two variants of magnet configuration offered for Metso's low-intensity drum separator for DMS applications. The two configurations (Fig. 6.13) demonstrate the trade-off between increasing magnetic loading capacity (Fig. 6.13(a)) to capture more particles and increasing field gradient (Fig. 6.13(b)) to capture

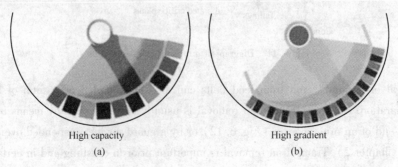

Fig. 6.13 Alternate magnet configurations for a wet drum separator
(a) High-capacity arrangement, and (b) high-gradient arrangement (Courtesy Metso)

finer or less susceptible particles. The high-capacity arrangement has fewer, larger poles, which results in a lower field gradient but a higher magnetic flux of 0.12T at a distance of 50mm from the roll surface (Metso, 2014a). The high gradient variant has more, smaller poles, resulting in a higher field gradient to better capture fine magnetic particles, but at reduced capacity (magnetic flux of only 0.06T at a distance of 50mm from the roll surface) (Metso, 2014a).

Although drum separators initially employed electromagnets, permanent magnets are used in modern devices, utilizing ceramic or rare earth magnetic alloys, which retain their intensity for an indefinite period (Norrgran and Marin, 1994). Separation in a drum separator occurs by the "pick-up" principle, wherein magnetic particles are lifted by the magnets and pinned to the drum and are conveyed out of the field, leaving the nonmagnetics (usually the gangue) in the tailings compartment. Water is introduced to provide flow, which keeps the pulp in suspension. Field intensities of up to 0.7T at the pole surfaces can be obtained in this type of separator.

The drum separators shown in Fig. 6.11 and Fig. 6.14 are of the concurrent type (as shown by the separator tank flow pattern in Fig. 6.14), whereby the concentrate is carried forward by the drum and passes through a gap, where it is compressed and dewatered before leaving the separator. This design is most effective for producing a clean magnetic concentrate from relatively coarse feeds (up to 6~8mm) and is widely used in dense medium recovery systems. In addition to the concurrent arrangement (Fig. 6.14), drum separators may also be configured with counter-current and counter-rotation arrangements (Fig. 6.15 and Fig. 6.16).

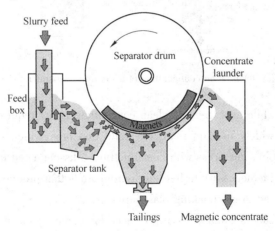

Fig. 6.14 Concurrent configuration of a wet drum separator (Courtesy Metso)

In a counter-current separator, the tailings are forced to travel in the opposite direction to the drum rotation and are discharged into the tailings chute. This type of separator is designed for finishing operations on relatively fine material, of particle size less than about 800μm. Pulp densities in this type of separator are typically lower than in the concurrent configuration.

The third possible configuration is the counter-rotation type, where the feed flows in the opposite direction to the rotation. This type is used in roughing operations, where occasional surges in feed must be handled, and where magnetic material losses are to be held to a minimum when high solids

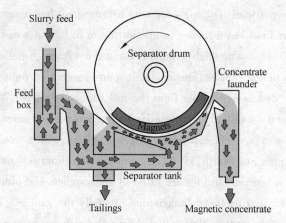

Fig. 6.15 Counter-current configuration of a wet drum separator (Courtesy Metso)

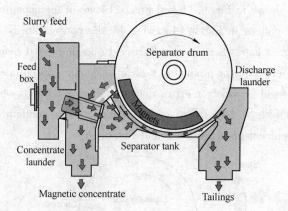

Fig. 6.16 Counter-rotation configuration of a wet drum separator (Courtesy Metso)

loading is encountered, while an extremely clean concentrate is not required.

Drum separators are widely used to treat low-grade iron ores, which contain 40% ~ 50% Fe, mainly as magnetite, but in some areas with hematite, finely disseminated in bands in hard siliceous rocks. Very fine grinding is necessary to free the iron minerals that produce a concentrate requiring pelletizing before being fed to steelmaking blast furnaces.

At the Iron Ore Company of Canada's Carol Project, low-intensity magnetic separation is used as an initial cobbing step on the cyclone overflow of the ball mill discharge to remove magnetite. This magnetite concentrate is then combined with the tailings of spiral gravity concentrators to be fed to a rougher wet drum low-intensity separator where magnetite is removed and sent directly to the pellet plant, with the tailings from the drum being sent for further gravity processing to concentrate any remaining hematite (Damjanovic and Goode, 2000).

The mechanism by which ferromagnetic particles are captured by low-intensity drum separators has been investigated for both high solids content (10% ~ 17% solids by weight) and low solids content (2%) (Rayner and Napier-Munn, 2000). For the high feed solids case, magnetic recovery occurs primarily via the formation of magnetic flocs, which are then captured. At lower feed solids, magnetic capture is not contingent on floc formation, as the increased distances between particles make it more difficult for flocs to form. These findings are significant as they provide information on the dominant variable affecting ferromagnetic losses to the tailings. At high feed solids content, particles with low magnetic susceptibility are lost to the tailings (higher magnetic susceptibility promotes floc formation), while at low feed solids content tailings losses are primarily fine ferromagnetic particles (Rayner and Napier-Munn, 2000).

At Palabora, the tailings from copper flotation are deslimed, after which the +105μm material is treated by wet low-intensity drum separators to recover 95% of the magnetite at a grade of 62% Fe.

Another example of wet low-intensity magnetic separation is the treatment of flotation tailings at the Niobec mine in Quebec, Canada. The Niobec mine employs multiple flotation stages to produce a pyrochlore concentrate including carbonate, pyrochlore, and sulfide flotation circuits. The deslimed tails from the carbonate flotation bank are fed to low-intensity drum magnetic separators to remove approximately 1t/h of magnetite assaying 68.30% Fe, 0.08% Nb_2O_5, 0.80% SiO_2, and 0.16% P_2O_5. The nonmagnetic product (containing the bulk of the Nb) is then sent to the pyrochlore flotation circuit for upgrading (Biss and Ayotte, 1989).

The cross-belt separator (Fig. 6.17) and disc separators, once widely used in the mineral sands industry, are now considered obsolete. They are being replaced with rare earth roll magnetic separators and rare earth drum magnetic separators (Arvidson, 2001).

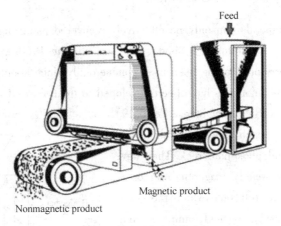

Fig. 6.17 Cross-belt separator

Rare earth roll separators use alternate magnetic and nonmagnetic laminations (like those illustrated in Fig. 6.6). Feed is carried onto the magnetic roll by a belt as shown in Fig. 6.18 to limit bouncing or scattering of particles and to ensure they all enter the magnetic zone with the same horizontal velocity. In a rare earth roll separator, the variables affecting separation are the magnetic field strength, the feed rate, the linear speed of the roll surface, and the particle size of the material (Eriez, 2003). Most importantly, the centrifugal force applied to the mineral particles by the roll surface must be optimized to achieve a sharp separation (Eriez, 2003). To control the centrifugal force, roll speed can be adjusted over a wide range, allowing the product quality to be "dialed in".

Dry rare earth drum separators provide a "fan" of separated particles, which can often be seen as distinct streams (Fig. 6.18). The fan can be separated into various grades of magnetic product and a nonmagnetic tailing. In some mineral sands applications, drum separators have been integrated with one or more rare earth rolls, arranged to treat the middlings particles from the drum. In any dry magnetic separator, the careful control of feed moisture is critical to avoid smaller particles sticking to larger particles (Oberteuffer, 1974). While increasing particle size increases the acceptable moisture limits, even at a particle size of 90% passing 20mm, the recommended moisture limit for Metso's dry drum separators is only 3% (Metso, 2014b).

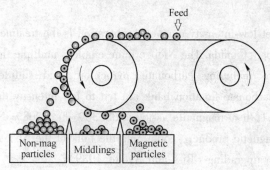

Fig. 6.18 Schematic of rare earth roll separator (Adapted from Dobbins et al. (2009))

6.5.2 High-Intensity Magnetic Separators

Weakly paramagnetic minerals can only be effectively recovered using high-intensity (B-fields of 2T or greater) magnetic separators (Svoboda, 1994). Until the 1960s, high-intensity separation was confined solely to dry ore, having been used commercially since about 1908. This is no longer the case, as many new technologies have been developed to treat slurried feeds.

Induced roll magnetic (IRM) separators (Fig. 6.19) are widely used to treat beach sands, wolframite and tin ores, glass sands, and phosphate rock. They have also been used to treat weakly magnetic iron ores, principally in Europe. The roll, onto which the ore is fed, is composed of phosphated steel laminates compressed together on a non-magnetic stainless steel shaft. By using two sizes of laminations, differing slightly in outer diameter, the roll is given a serrated profile, which promotes the high field intensity and gradient required. Field strengths of up to 2.2T are attainable in the gap between feed pole and roll. Nonmagnetic particles are thrown off the roll into the tailings compartment, whereas magnetics are held, carried out of the influence of the field and deposited into the magnetics compartment. The

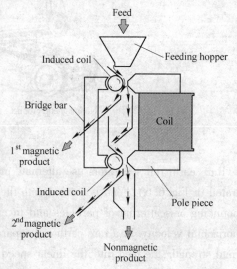

Fig. 6.19 Schematic of an induced roll separator

gap between the feed pole and rotor is adjustable and is usually decreased from pole to pole (to create a higher effective magnetic field strength) to take off successively more weakly magnetic products.

The primary variables affecting separation using an IRM separator are the magnetic susceptibility of the mineral particles, the applied magnetic field intensity, the size of the particles, and the speed of the roll (Singh et al.,2013). The setting of the splitter plates cutting into the trajectory of the discharged material is also of importance.

In most cases, IRM separators have been replaced by the more recently developed (circa 1980) rare earth drum and roll separators, which are capable of field intensities of up to 0.7T and 2.1T, respectively (Norrgran and Marin, 1994). The advantages of rare earth roll separators over IRM separators include: lower operating costs due to decreased energy requirements, less weight leading to lower construction and installation costs, higher throughput, fewer required stages, and increased flexibility in roll configuration which allows for improved separation at various size ranges (Dobbins and Sherrell, 2010).

Dry high-intensity separation is largely restricted to ores containing little, if any, material finer than about 75μm. The effectiveness of separation on such fine material is severely reduced by the effects of air currents, particleparticle adhesion, and particlerotor adhesion.

Without doubt, the greatest advance in the field of magnetic separation was the development of continuous WHIMSs (Lawver and Hopstock, 1974). These devices have reduced the minimum particle size for efficient magnetic separation compared to dry high-intensity methods. In some flowsheets, expensive drying operations, necessary prior to a dry separation, can be eliminated by using an entirely wet concentration system.

Perhaps the most well-known WHIMS machine is the Jones separator, the design principle of which is utilized in many other types of wet separators found today. The machine has a strong main frame (Fig. 6.20(a)) made of structural steel. The magnet yokes are welded to this frame, with the electromagnetic coils enclosed in air-cooled cases. The separation takes place in the plate boxes, which are on the periphery of the one or two rotors attached to the central roller shaft and carried into and out of the magnetic field in a carousel (Fig. 6.20(b)). The feed, which is thoroughly mixed slurry, flows through the plate boxes via fitted pipes and launders into the plate boxes (Fig. 6.21),

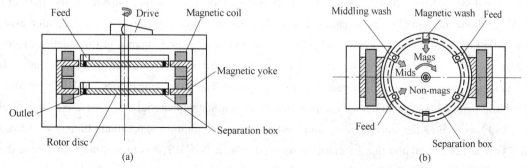

Fig. 6.20 The Jones high-intensity wet magnetic separator in cross section
(a) Plan view, and (b) top view

which are grooved to concentrate the magnetic field at the tip of the ridges. Feeding is continuous due to the rotation of the plate boxes on the rotors and the feed points are at the leading edges of the magnetic fields (Fig. 6.20(b)). Each rotor has two feed points diametrically opposed to one another.

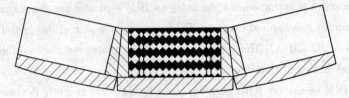

Fig. 6.21 Jones plate box showing grooved plates and spacer bars

The weakly magnetic particles are held by the plates, whereas the remaining nonmagnetic particle slurry passes through the plate boxes and is collected in a launder. Before leaving the field any entrained nonmagnetics are washed out by low-pressure water and are collected as a middlings product.

When the plate boxes reach a point midway between the two magnetic poles, where the magnetic field is essentially zero, the magnetic particles are washed out using high-pressure scour water sprays operating at up to 5bar. Field intensities of over 2T can be produced in these machines, although the applied magnetic field strength should be carefully selected depending on the application (see Section 6.4.2). The production of a 1.5T field requires electric power consumption in the coils of 16kW per pole.

There are currently two types of WHIMS machines, one that uses electromagnetic coils to generate the required field strength, and the other that employs rare earth permanent magnets. They are used in different applications; the weaker magnetic field strength produced by rare earth permanent magnets may be insufficient to concentrate some weakly paramagnetic minerals. The variables to consider before installing a traditional horizontal carousel WHIMS include: the feed characteristics (slurry density, feed rate, particle size, magnetic susceptibility of the target magnetic mineral), the product requirements (volume of solids to be removed, required grade of products), and the cost of power (Eriez, 2008). From these considerations the design and operation of the separator can be tailored by changing the following: the magnetic field intensity and/or configuration, the speed of the carousel, the setting of the middling splitter, the pressure/volume of wash water, and the type of matrix material (Eriez, 2008). The selection of matrix type has a direct impact on the magnetic field gradient present in the separation chamber. As explained in Section 6.4.2, increasing magnetic field can in some applications actually cause decreased performance of the magnetic separation step and it is for this reason that improvements in the separation of paramagnetic materials focus largely on achieving a high magnetic field gradient. The Eriez model SSS-I WHIMS employs the basic principles of WHIMS with improvements in the matrix material (to generate a high field gradient) as well as the slurry feeding and washing steps (to improve separation efficiency) (Eriez and Gzrinm, 2014). While this separator is referred to as a WHIMS, it is in fact more similar to the SLon VPHGMS mentioned in Sections 6.4.1 and 6.5.3. Further discussion on high-gradient magnetic separation (HGMS) may be found in Section 6.5.3.

Wet high-intensity magnetic separation has its greatest use in the concentration of low-grade iron ores containing hematite, where they are an alternative to flotation or gravity methods. The decision to select magnetic separation for the concentration of hematite from iron ore must balance the relative ease with which hematite may be concentrated in such a separator against the high capital cost of such separators. It has been shown by White (1978) that the capital cost of flotation equipment for concentrating weakly magnetic ore is about 20% that of a Jones separator installation, although flotation operating costs are about three times higher (and may be even higher if water treatment is required). Total cost depends on terms for capital depreciation; over 10 years or longer the high-intensity magnetic separator may be more attractive than flotation.

In addition to recovery of hematite (and other iron oxides such as goethite), wet high-intensity separators are now in operation for a wide range of duties, including removal of magnetic impurities from cassiterite concentrates, removal of fine magnetic material from asbestos, removal of iron oxides and ferrosilicate minerals from industrial minerals such as quartz and clay, concentration of ilmenite, wolframite, and chromite, removal of magnetic impurities from scheelite concentrates, purification of talc, the recovery of non-sulfide molybdenum-bearing minerals from flotation tailings, and the removal of Fe-oxides and FeTi-oxides from zircon and rutile in heavy mineral beach sands (Corrans and Svoboda, 1985; Eriez, 2008). In the PGM-bearing Merensky Reef (South Africa), WHIMS has been used to remove much of the strongly paramagnetic orthopyroxene gangue from the PGM-containing chromite (Corrans and Svoboda, 1985). WHIMS has also been successfully used for the recovery of gold and uranium from cyanidation residues in South Africa (Corrans, 1984). Magnetic separation can be used to recover some of the free gold, and much of the silicate-locked gold, due to the presence of iron impurities and coatings. In the case of uranium leaching, small amounts of iron (from milling) may act as reducing agents and negatively affect the oxidation of U^{4+} to U^{6+}; treatment via WHIMS can reduce the consumption of oxidizing agents by removing a large portion of this iron prior to leaching (Corrans and Svoboda, 1985).

At the Cliffs Wabush iron ore mine in Labrador, Canada (Fig. 6.22), the cyclone overflow from the tailings of a rougher spiral bank is sent to a magnetic scavenger circuit utilizing both low-intensity drum separation and WHIMS. This circuit employs the low-intensity (0.07T) drum separators to remove fine magnetite particles lost during the spiral gravity concentration step, followed by a WHIMS step using 100t/h Jones separators which are operated at field strengths of 1T to concentrate fine hematite. Cleaning of only the gravity tailings by magnetic separation is preferred, as relatively small amounts of magnetic concentrate have to be handled, the bulk of the material being essentially unaffected by the magnetic field. The concentrate produced from this magnetic scavenging step is eventually recombined with the spiral concentrate before feeding to the pelletizing plant (Damjanovic and Goode, 2000).

The paramagnetic properties of some sulfide minerals, such as chalcopyrite and marmatite (high Fe form of sphalerite), have been exploited by applying wet high-intensity magnetic separation to augment differential flotation processes (Tawil and Morales, 1985). Testwork showed that a Chilean copper concentrate could be upgraded from 23.8% to 30.2% Cu, at 87% recovery.

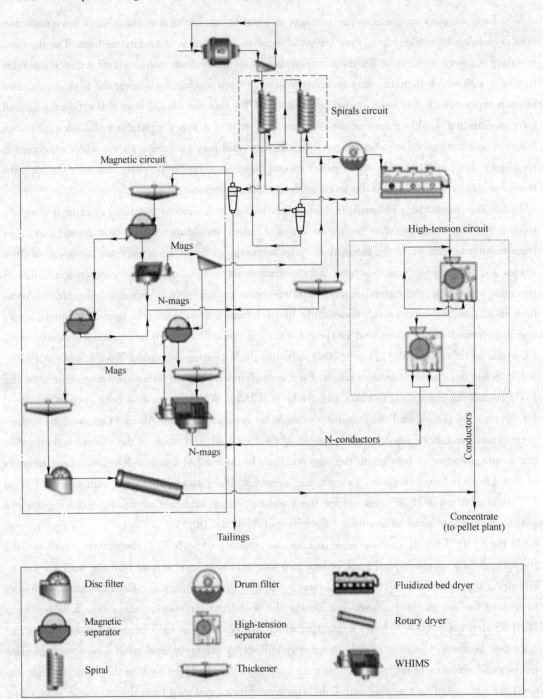

Fig. 6.22 Flowsheet from Cliffs-Wabush iron ore mine
(Adapted from Damjanovic and Goode (2000))

6.5.3 High-Gradient Magnetic Separators

As noted in Eq. (6.7), in order to separate paramagnetic minerals of low magnetic susceptibility and/or fine size, high field gradients are required. These are generated by exploiting the ferromagnetic properties of iron to generate a high B-field (induced field) many hundreds of times greater than the applied H-field. This, however, requires that the iron be in the volume where separation takes place. The steel plates in a Jones separator, for example, occupy up to 60% of the process volume. Thus, high-intensity magnetic separators using conventional iron circuits tend to be very massive and heavy in relation to their capacity. A large separator may contain over 200t of iron to carry the flux, hence capital and installation costs are high.

Instead of using one large convergent field in the gap of a magnetic circuit, as in the Jones separator, in HGMS a solenoid is used to generate a uniform field with a solenoid core, or working volume, filled with a matrix of secondary ferromagnetic poles, such as ball bearings, or wire wool, the latter filling only about 10% of the working volume. Each secondary pole, due to its high permeability, can produce maximum field strengths of 2T at their surface, but more importantly, each pole produces, in its immediate vicinity, high field gradients of up to 14T/mm. Thus, a multitude of high gradients across numerous small gaps, centered on each of the secondary poles, is achieved.

The solenoid can be clad externally with an iron frame to form a continuous return path for the magnetic flux, thus reducing the energy consumption for driving the coil by a factor of about 2. The matrix is held in a canister into which the slurry is fed. Both continuous and batch-type HGMS are available, with batch-type HGMS requiring periodic demagnetization in order to remove accumulated magnetic particles, while the continuous HGMS (Fig. 6. 23) operates in a carousel-type configuration similar to the Jones WHIMS (Metso, 2014c, d).

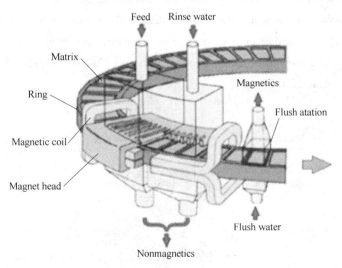

Fig. 6. 23 Metso HGMS operating principle(Courtesy Metso)

An inherent disadvantage of high-gradient separators is that in producing an increase in field gradient, the working gap between secondary poles is reduced, so that the magnetic force has a short reach of no more than about 1mm. It is therefore necessary to use gaps of only about 2mm between poles, such that the matrix separators are best suited to the treatment of very fine particles. They are used mainly in the kaolin industry for removing iron-containing particles which lower brightness.

In order to address some of the deficiencies in the design of HGMS, new horizontally fed vertical carousel separators have been designed that incorporate a pulsating feed system to ensure particle dispersion (i.e., avoid flocculation) and prevent nonmagnetic entrainment. The SLon VPHGMS (Fig. 6.24 and Fig. 6.25) employs a unique matrix of steel rods oriented perpendicular to the applied magnetic field (Section 6.4.1) as well as flushing of trapped magnetic particles (Fig. 6.26) in the reverse direction to the feed in order to reduce particle momentum, maximize particle trapping, and improve separation (Outotec, 2013). The rod diameter in the matrix may be tailored for the given application to vary the maximum particle size that can pass through the separator from 0.6mm up to 3.0mm (Outotec, 2013). The averaged magnetic field intensity across the entire VPHGMS is no greater than 1.3T; however, as the steel rod matrix becomes saturated, intensities up to 1.8T can be achieved at the matrix surface with an applied magnetic field of only 1T (Outotec, 2013). The SLon separator has been applied in the concentration of fine particles such as hematite and ilmenite, and for desulfurization and dephosporization of iron ore feeds prior to steelmaking (Xiong, 1994, 2004). Eriez also offers a vertical carousel-type WHIMS with similar innovations to the SLon VPHGMS, such as pulsating feed and high capacity due to improved matrix washing (Eriez and Gzrinm, 2014). The recently developed version of the Eriez separator is somewhat confusingly referred to as WHIMS, while using a rod matrix to produce high magnetic field gradients in a manner similar to HGMS (Eriez and Gzrinm, 2014). The Eriez separator has been successfully applied to the following: the concentration of Fe-bearing minerals (hematite, limonite, siderite, chromite), the cleaning of nonferrous ores (quartz, cassiterite, garnet), the recovery of rare earth minerals, and the purification of nonmetallic ores (quartz, feldspar, kaolin, alusite, kyanite) (Eriez and Gzrinm, 2014).

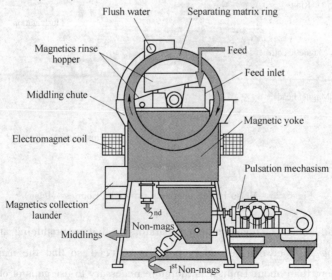

Fig. 6.24   SLon VPHGMS (Courtesy Outotec)

6.5 TYPES OF MAGNETIC SEPARATOR

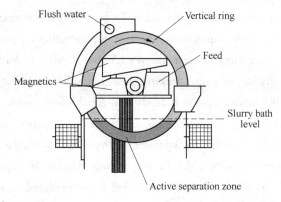

Fig. 6.25 Plan view of separation zone in SLon VPHGMS (Courtesy Outotec)

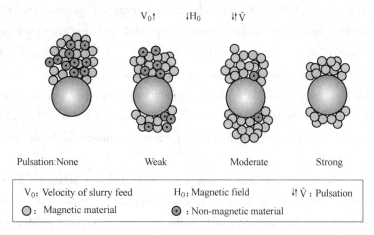

Fig. 6.26 Build up on the matrix in the SLon separator (Courtesy Outotec)

6.5.4 Superconducting Separators

Future developments and applications of magnetic separation in the mineral industry lie in the creation and use of increasingly higher product of field and field gradient, that is, the "force factor". Matrix separators with very high field gradients and multiple small working gaps can draw little benefit from field strengths above the saturation levels of the secondary poles (~ 2T for an iron/steel matrix material). As discussed in Section 6.4.1, the alternative to HGMS is OGMS, where separators with large working volumes deflect coarser particles at high capacity, rather than capture particles, as in HGMS. As the gradient in OGMS is relatively low, these separators need to use the highest possible field strengths to generate the high magnetic forces required to treat weakly paramagnetic particles. Field strengths in excess of 2T can only be generated economically by the use of superconducting magnets (Kopp, 1991; Watson, 1994).

Certain alloys have the property of presenting no resistance to electric currents at extremely low temperatures. An example is niobiumtitanium at 4.2K, the temperature of liquid helium. Once a current is established through a coil made from a superconducting material, it will continue to flow without being connected to a power source, and the coil will become, in effect, a permanent magnet. Superconducting magnets can produce extremely intense and uniform magnetic fields, of up to

15T. The main problem, of course, is in maintaining the extremely low temperatures. In 1986, a Ba/La/Cu oxide composite was made superconductive at 35K, promoting a race to prepare ceramic oxides with much higher superconducting temperatures (Malati, 1990). Unfortunately, these materials are of a highly complex crystal structure, making them difficult to fabricate into wires. They also have a low current-carrying capacity, so it is likely that for the foreseeable future superconducting magnets will be made from ductile niobium alloys, embedded in a copper matrix.

The main advantage of superconducting separators is that elevated magnetic field strength increases the maximum feed slurry velocity with a corresponding increase in capacity (Kopp, 1991). In order to fully utilize this capacity, downtime for removal of accumulated magnetic particles from the working volume of the separator must be minimized through the use of a reciprocating or continuously cycling matrix (Kopp, 1991). Another advantage of these separators is the reduced weight of the separators (smaller coils and windings along with much less iron required compared to the heavy frames and matrix materials used in HGMS) (Gillet and Diot, 1999). The factors limiting the adoption of superconducting separators are the difficulties in maintaining the very low temperatures necessary for the material to retain its superconducting properties against heat leaks, and the high energy costs associated with maintaining this refrigeration (Kopp, 1991). Superconducting magnets are generally only viable when large field volumes and magnetic fields greater than 2T are required (Kopp, 1991).

In 1986, a superconducting HGMS was designed and built by Eriez Magnetics to remove magnetic (and colored) contaminants from kaolinite clay for operations in the United States. This machine used only about 0.007kW in producing 5T of flux, the ancillary equipment needed requiring another 20kW. In comparison, a conventional 2T high-gradient separator of similar throughput would need about 250kW to produce the flux, and at least another 30kW to cool the magnet windings.

The 5T machine is an assembly of concentric components (Fig. 6.27). A removable processing

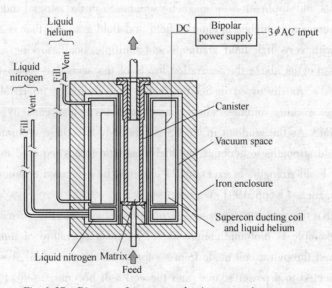

Fig. 6.27 Diagram of a superconducting magnetic separator

canister is installed in a processing chamber located at the center of the assembly. This is surrounded by a double-walled, vacuum-insulated container that accommodates the superconductive niobium/titanium-tantalum winding and the liquid helium coolant. A thermal shield, cooled with liquid nitrogen to 77K, limits radiation into the cryostat. In operation, the supply of slurry is periodically cut off, the magnetic field is shut down, and the canister backwashed with water to clear out accumulated magnetic contaminants.

A picture of a superconducting magnetic separator in a horizontal arrangement installed in a plant is shown in Fig. 6.28.

Fig. 6.28 Superconducting magnetic separator(Courtesy Imerys)

An open-gradient drum magnetic separator with a superconducting magnet system has been operating commercially since the 1980s (Unkelbach and Kellerwessel, 1985; Wasmuth and Unkelbach, 1991). Although separation is identical to that in conventional drum separators, the magnetic flux density at the drum surface can reach over 4T.

The development of HGMS and superconducting separators capable of concentrating very fine or very weakly magnetic mineral particles has prompted the application of magnetic separation techniques to treat many waste streams from mineral processing operations. Fine ($<10\mu m$), weakly magnetic hematite and limonite have been recovered by a combination of selective flocculation using sodium oleate and kerosene followed by HGMS (Song et al., 2002). HGMS has been used to recover fine gold-bearing leach residues from uranium processing, and fine Pb minerals containing V and Zn from a mining waste dump (Watson and Beharrell, 2006). A single-stage extraction of ilmenite from highly magnetic gangue minerals has been developed using a super-conducting HGMS system (the difference in magnetic susceptibility between ilmenite and gangue is only significant at very high magnetic field strength). However, this process is still faced with the typical challenges associated with an industrial installation of a superconducting separator (Watson and Beharrell, 2006). Another interesting, and potentially significant, application of HGMS is in the treatment of wastewater streams containing heavy metal ions. Multiple authors have developed processes where the metal ions to be removed are coprecipitated with Fe ions to form a fine, dispersed magnetite phase which can be easily extracted through the use of HGMS (Gillet et al., 1999; Karapinar, 2003).

6.6 ELECTRICAL SEPARATION

Electrical separation exploits the differences in electrical conductivity between different minerals in a feed. Since almost all minerals show some difference in conductivity, it would appear to represent the universal concentrating method. In practice, however, the method has fairly limited applications due to the required processing conditions (notably a perfectly dry feed), and its greatest use is in separating some of the minerals found in heavy mineral sands from beach or stream placer deposits (Dance and Morrison, 1992). Electrical separation also suffers from a similar disadvantage to dry magnetic separation the capacity is very small for finely divided material. For most efficient operation, the feed to most electrical separators should be in a layer, one particle deep, which reduces the throughput if the particle size is small ($<75\mu m$).

There are two distinct forces which may be considered in the context of electrical separation. The electrophoretic force is the force experienced by a charged particle under the influence of an electric field, and the dielectrophoretic force is the force experienced by a neutral particle in a fluid when subjected to a nonuniform electric field. The dielectrophoretic force is somewhat analogous to magnetic force as it relies on the polarization of a neutral particle into an electric dipole as well as a nonuniform applied field (Lockhart, 1984). The deliberate use of die-lectrophoresis is almost nonexistent in mineral processing however, as the electrophoretic force is much stronger (Lockhart, 1984).

In order to exploit the electrophoretic force for mineral separation, a treatment step prior to separation is required in all electrical separators to selectively charge the mineral particles. This selective development of charges on particles relies on conductivity differences between the minerals. As most electrical conduction occurs in the surface layers of atoms (Dance and Morrison, 1992), electrical separation may be thought of as a surface-based separation, similar to flotation, as opposed to magnetic and gravity separation which rely on differences in bulk properties (magnetic susceptibility, specific gravity).

There are three main mechanisms by which minerals are charged: ion bombardment (corona charging), conductive induction, and frictional charging (tribocharging or contact electrification). Each of these three mechanisms has a corresponding separator type, the details of which are described in the following sections. To understand electrical separation methods, knowledge of the electrical properties of materials is required. Introduction to the relevant concepts, as they apply to mineral processing, along with detailed descriptions of many industrial separators, may be found in the comprehensive reviews by Kelly and Spottiswood (1989 ac) and Manouchehri et al. (2000).

6.6.1 Ion Bombardment

Charging via ion bombardment occurs as a high voltage is applied between two electrodes so that the gas near the electrodes ionizes and forms a corona discharge, a continuous flow of gaseous ions. Mineral particles passing through this corona are bombarded with the flow of ions and develop a charge. A similar mechanism of charge application is employed in electrostatic precipitators used to remove fine particulate matter from flowing gas streams. In mineral separation applications, different conductivities of the charged mineral particles then result in different rates of charge decay and correspondingly different forces experienced by the particles.

The typical separator relying on corona charging is the high-tension roll (HTR) separator (Fig. 6.29). In this separator the feed, a mixture of ore minerals of varying susceptibilities to surface charging, is fed to a rotating drum made from mild steel, or some other conducting material, which is grounded through its support bearings. An electrode assembly, comprising a brass tube in front of which is supported a length of fine wire, spans the complete length of the roll and is supplied with a fully rectified DC supply of up to 50kV, usually of negative polarity. Together these two electrodes act to create a dense high-voltage discharge. The fine wire tends to discharge readily, whereas the large electrode tends to have a short range, dense, nondischarging field. This combination creates a strong discharge pattern that may be "beamed" in a definite direction and concentrated to a very narrow arc. The voltage supplied should be such that ionization of the air takes place. Arcing between the electrode and the roll must be avoided, as this destroys the ionization.

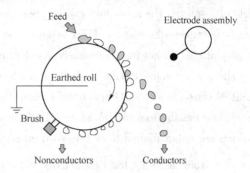

Fig. 6.29 Diagram of HTR separator

When ionization occurs, the mineral particles receive a spray discharge of ions which gives all particles in the corona field a surface charge. As the HTR drum rotates and particles are moved outside of the corona field, weakly conductive particles maintain a high surface charge, causing them to be attracted to and pinned to the rotor surface. This is often referred to as pinning by the image force (Fig. 6.30), and it may be explained by the charged mineral particle inducing a charge of opposite sign on the rotor (Dance and Morrison, 1992). Pinned particles are removed from the rotor surface either through the eventual decay of their surface charge or mechanically by means of a brush.

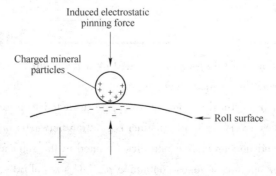

Fig. 6.30 Representation of pinning force experienced by nonconducting particle in a HTR separator (Courtesy OreKinetics)

Particles of relatively high conductivity lose their surface charge as the charge rapidly dissipates to the earthed rotor. The centrifugal force of the rotor, along with gravitational and frictional forces, is then able to throw these particles from the roll and away from the relatively low-conductivity particles that remain pinned so that two streams of particles develop which may be collected separately through the use of a splitter. The separation can be optimized by varying the splitter position. However, predicting particle trajectories from an HTR separator is challenging, as Edward et al. (1995) have shown that particles do not instantaneously accelerate to the roll speed, due to slip on the rotor surface.

The primary industrial use of HTR separators is in the processing of heavy mineral sands (Dance and Morrison, 1992). Other uses include coal cleaning (Butcher and Rowson, 1995), and recycling metals from plastic waste (Dascalescu et al., 1993). Table 6.2 shows typical minerals which are either pinned to or thrown from the rotor during HTR separation.

Table 6.2 Typical Conductivity and Behavior of Minerals in a High-Tension Separator

Nonconductive minerals (Pinned)	Conductive minerals (Thrown)
Apatite	Cassiterite
Barite	Chromite
Calcite	Diamond
Coal	Feldspar
Corundum	Galena
Garnet	Gold
Gypsum	Hematite
Kyanite	Ilmenite
Monazite	Limonite
Quartz	Magnetite
Scheelite	Pyrite
Sillimonite	Rutile
Spinel	Sphalerite
Tourmaline	Stibnite
Zircon	Tantalite
	Wolframite

A combination of pinning and lifting can be created by using a third "static" electrode following the corona discharge electrodes with a diameter large enough to preclude corona discharge. The conducting particles, which are thrown from the rotor, are attracted to this third electrode, and the combined process produces a very wide and distinct separation between conductive particles (lifted from the rotor surface) and nonconductive particles (pinned to the rotor surface).

High-tension separators operate on feeds containing particle sizes of $60 \sim 500 \mu m$. Particle size influences separation behavior, as surface charges on a coarse particle are lower in relation to its mass than on a fine particle. Thus, a coarse particle is more readily thrown from the roll surface, and the conducting fraction (particles thrown from the rotor) often contains a small proportion of coarse nonconductors. Similarly, the finer the particles the more they are influenced by the surface charge, and the nonconducting fraction often contains some fine conducting particles. This cross-contamination may also be interpreted in terms of the interplay between the centrifugal force on a particle and the image force acting to pin a charged particle to a grounded surface. The centrifugal force varies with particle mass, while the image force varies with surface area (as charge is accumulated on the particle surface) so, consequently, the centrifugal force is dominant at coarse particle sizes (Dance and Morrison, 1992; Svoboda, 1993).

Some machine factors affecting the operation of an HTR separator include: geometry of the electrode assembly, electrode voltage and polarity, rotor speed, rotor diameter, and splitter position (Dance and Morrison, 1992). Larger rotor diameters help to increase recovery, while a smaller rotor diameter improves the grade of the conducting fraction (Svoboda, 1993). A similar dependence exists for particle density, rotor speed, and the coefficient of friction between the particle and rotor surface, so that separation selectivity is maximized at low particle density, small rotor diameter, high rotor speed, and high coefficient of friction (Svoboda, 1993). The effect of rotor speed on separation is complex and dependent on the conductivity of a given particle, as the act of increasing rotor speed decreases the time available for charge decay. In this way, increased rotor speed increases the chance that a conductive mineral particle will report to the nonconductor fraction, while high rotor speeds will also increase the centrifugal force on a nonconductive particle so that it is more likely to incorrectly report to the conductive fraction (Svoboda, 1993). Stated another way, increased rotor speed simultaneously increases the minimum particle size necessary for a conductive particle to be thrown from the rotor while decreasing the maximum nonconductive particle size that will be pinned to the rotor (Svoboda, 1993).

While HTR separation primarily exploits the differences in conductivities between minerals, an equally important criteria for successful operation is the presence of at least one strongly conductive (on an absolute basis) mineral species in the separator feed. It has been shown by Svoboda (1993) that very large differences in mineral conductivities (up to an order of magnitude) will not result in a sharp separation if both minerals are weak conductors. Conversely, two strongly conducting minerals can be separated with only a small difference in their conductivities.

HTR separators have been one of the mainstays of the mineral sands industry for decades. Very little development of the machines has occurred in that period; their generally poor single pass separation has been tolerated, and overcome by using multiple machines and multiple recycle streams. However, in the last few years innovative new designs have started to appear, from new as well as established manufacturers. Roche Mining (MT) have developed the Carara HTR separator, which incorporates an additional insulated plate static electrode to help deflect the path of conductive particles thrown from the rotor (Germain et al., 2003). Outokumpu Technology developed the eForce HTR separator, which also incorporates additional static electrodes, as well as an electrostatic feed classifier (Elder and Yan, 2003).

Ore Kinetics has introduced the new Corona Stat machine (Fig. 6.31), which is a significant improvement on existing HTR designs as it employs additional static electrodes to improve the efficiency of separation. Unlike existing machines, the static electrodes are not exposed, making the machines much safer to operate. The key improvement in the Corona Stat design relative to traditional HTR separators is the presence of induction electrodes, which simultaneously increase the pinning force on nonconducting particles and increase the rate of charge decay for conductive particles (Fig. 6. 32). This results in a larger distance between the two particle streams and therefore an improved separation.

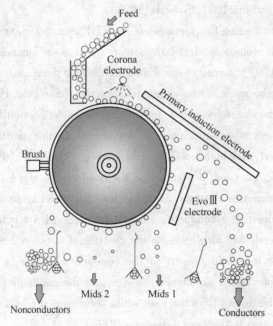

Fig. 6. 31 Diagram of Corona Stat separator (Courtesy Ore Kinetics)

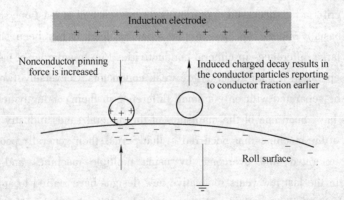

Fig. 6. 32 Effect of induction electrode in Corona Stat separator (Courtesy Ore Kinetics)

6.6.2 Conductive Induction

The second charging mechanism used in electrical separators is conductive induction, in which polarization of a mineral particle occurs upon exposure to an electric field. Similar to charge decay in

HTR separators, the ability of the mineral particle to respond to this induced polarization is directly related to its conductivity. Polarization results when an uncharged particle develops an opposite charge, relative to the electrode creating the electric field, at the surface closest to the electrode and a corresponding like charge to the electrode on the particle surface furthest from the electrode. Conductive particles are able to redistribute these induced charges across the particle surface, while nonconductive particles are unable to redistribute these charges and will remain polarized. The electric force on a polarized particle is a function of the degree to which it polarizes, which is in turn affected by both the size and shape of the particle (Manouchehri et al. ,2000). When a polarized particle contacts a conductive surface it may conduct charge of one polarity to the surface, leaving a net charge on the particle. In such a situation nonconductive particles (with no net charge) will experience no attraction from an applied electric field, whereas conductive particles will be attracted to an oppositely charged electrode (Kelly and Spottiswood, 1989a). Conductive induction can therefore be thought of as a process in which charges are induced on uncharged conductive particles, leaving nonconductive particles with no net charge. A graphical representation of conductive induction may be seen in Fig. 6. 33.

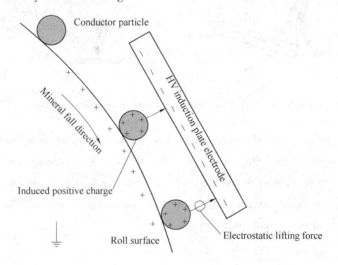

Fig. 6. 33 Representation of conductive induction(Courtesy Ore Kinetics)

Separators exploiting this charging mechanism are typically used to separate strongly conductive particles from weakly conductive particles and employ static electrodes to "lift" charged conductive particles from a grounded surface while nonconductive particles remain pinned to that surface. The most common such separator is the electrostatic plate (ESP) separator. In an ESP separator, material is gravity fed through the separator and the force on the charged particles acts to counteract the force of gravity. In contrast to HTR separators, coarse particles will tend to report to the nonconducting fraction, which is why final cleaning of the products of HTR separation is often carried out in purely electrostatic separators.

Modern electrostatic separators are of the plate or screen type (Fig. 6. 34), the former being used to clean small amounts of nonconductors from a predominantly conducting feed (Fig. 6. 34(a)), while the screen separators remove small amounts of conductors from a mainly nonconducting feed (Fig. 6.34(b)). The principle of operation is the same for both types of separator. The feed particles gravitate down a sloping, grounded plate into an electrostatic field induced by a large, oval-shaped, and high-voltage electrode. Fine particles are most affected by the lifting force, and so fine conductive particles are preferentially lifted to the electrode, whereas coarse nonconductors are most efficiently rejected. Machine parameters affecting ESP separators include: electrode geometry, electrode voltage and polarity, plate curvature, and position of the splitters (Dance and Morrison, 1992). For both HTR and ESP separators, system humidity is intentionally kept low, as excess moisture may alter the conductivity of the fluid medium of the separator (the air) as well as affecting the conductivity of the particle surface through the dual effects of water molecules themselves and dissolved ions in the water (Kelly and Spottiswood, 1989b).

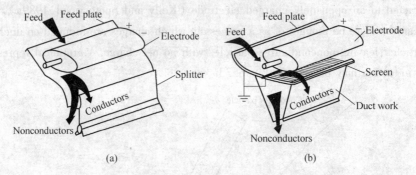

Fig. 6.34 (a) Plate, and (b) screen electrostatic separators

Similar to the Corona Stat for HTR separation, Ore Kinetics has also developed an improved ESP separator known as the Ultra Stat separator (Fig. 6.35). The primary improvements in this separator are different geometries of the electrode and particle feed path, the presence of secondary induction electrodes to further increase the lifting force on charged conductive particles as well as a secondary roll to clean the primary roll surface.

6.6.3 Triboelectric Charging

The final charging mechanism used in mineral processing is triboelectrification, or contact electrification, in which two materials of dissimilar electrical properties exchange electrons upon coming into contact with one another. As most minerals are semi-conductors, with volume conductivities between 10^{-5} and $10^{-8} \Omega/m$ (Manouchehri et al., 2000), the charge acquired by two minerals after contacting one another may be predicted by the relative Fermi levels (energy level at which 50% of the energy states in a material are occupied by electrons) of the two minerals (Kelly and Spottiswood, 1989c). An alternative measure also used to predict triboelectric charging behavior is the work function of a material, which is a measure of the energy required to bring an electron from the Fermi level of a given material to a free electron state (Kelly and Spottiswood, 1989c). A mineral with a low Fermi level must therefore have a higher work function than a mineral with a higher Fermi level. When two mineral particles come into frictional contact their Fermi levels will equalize,

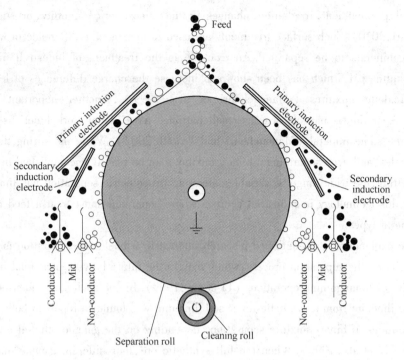

Fig. 6.35 Diagram of Ultra Stat separator (Courtesy Ore Kinetics)

with the mineral with the highest Fermi level losing electrons to the mineral with the lower Fermi level (Manouchehri et al.,2000). The mineral with the highest work function (lower Fermi level) becomes negatively charged and the opposing mineral becomes positively charged. The potential applications of triboelectric separation are immense, as separation does not require large differences in mineral conductivity and virtually every binary mixture of minerals will possess a difference in work function.

Once the minerals have acquired a charge, they are often separated using a free-fall design consisting of two charged electrodes which deflect mineral particles based on their surface charge with the mineral particles collected in different bins. Such separators have been used on a lab scale to separate quartz from wollastonite and calcite, calcite from insoluble silicates (Manouchehri and Fawell,2002), and on an industrial scale to beneficiate potash (Lockhart,1984).

In all triboelectric charging devices, mineral particles come into contact with not just one another, but also the material from which the conveying device is constructed. It is therefore important to take this into consideration, as different materials, such as brass and Teflon, have large differences in work function and will therefore produce corresponding differences in the charges produced on a given mineral particle (Dwari et al.,2009). Even if the charge induced on a mineral surface using different charging materials is of the same sign, the amount of the charge transfer between two materials is dependent on the differences between the Fermi levels of the two materials.

Charge acquisition of mineral particles in triboelectric separation may also be controlled through the use of a surface treatment prior to tribocharging such as: surface cleaning, chemical pretreatment, thermal pretreatment, irradiation, changes in the atmospheric humidity, or surface doping (Manouchehri, 2010). Such surface treatments are used to increase the difference in work function between two minerals to be separated. An example is the treatment of industrial minerals with H_3BO_3 at alkaline pH, which has been shown to increase the charge differences of feldsparquartz and feldsparcalcite mixtures (Manouchehri et al., 1999). Another important variable in triboelectric separation is particle size, as small particles have higher work functions than coarse particles of the same mineral (Manouchehri and Fawell, 2002). While measuring the charge on mineral particles, and even separating a binary mixture, can be readily accomplished in a controlled laboratory setting, the wide range of variables affecting triboelectric separation has limited the applications of this technology in industrial settings where separators must treat a feed consisting of multiple mineral types.

One of the explanations for continued research interest in triboelectric separation is the minimal effect of gravity on the separation process, which may in the future be very beneficial in developing extraterrestrial or lunar mining operations (Li et al., 1999). In one such study, focused on the beneficiation of ilmenite from a synthetic lunar soil, ilmenite was found to report to both positive and negative electrodes in binary mixture separations depending on the gangue mineral chosen for the feed mixture (Li et al., 1999). When the full synthetic ore, four different gangue minerals along with the ilmenite, was processed through the triboelectric separation unit ilmenite was found to be concentrated (by a factor of 2~3) in the neutral particle collection bin, evidently acquiring little net charge due to the presence of gangue minerals with both higher and lower work functions than the ilmenite (Li et al., 1999). This finding is illustrative of the inherent difficulties in predicting mineral behavior through triboelectric separations in an industrial setting.

Recently a new triboelectric separator, the ST separator, has been developed by Separation Technologies which employs conventional interparticle contact to tri-bocharge mineral particles and a continuous loop open-mesh belt that travels at high speeds (5~20m/s) between positive and negative electrodes for particle separation (Fig. 6.36) (Bittner et al., 2014). Feed enters from the top of the unit (at feed rates of up to 40t/h) with positive and negative charged particles exiting from opposite ends of the separator (Fig. 6.37). The separation occurs within a narrow gap (<1.5cm) between the electrodes. The top and bottom sections of the belt move in opposite directions, setting up a high particle number density, counter current flow within the electrode gap. Particles must travel across only a small fraction of the electrode gap (across the zone of high shear and lower velocities) under electrostatic forces to be separated into the oppositely flowing streams (Fig. 6.38). The counter current highly turbulent flow enables multiple stages of separation to occur within a single pass through the separator, increasing both grade and recovery of the product streams (Bittner et al., 2014). This multistage separation zone requires that the particles maintain their charge, which is made possible due to the high degree of interparticle contacts occurring throughout the separation zone (Bittner et al., 2014). This separator can process particles from 1 to 300μm, which is much smaller than conventional free-fall and HTR separators. It has been widely employed industrially in removing unburned coal char from fly ash (10~20μm median diameter) generated by coal-fired

6.6 ELECTRICAL SEPARATION

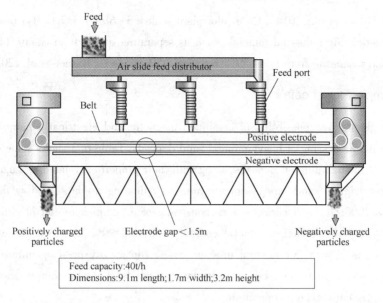

Fig. 6.36 Diagram of ST separator (Adapted from Bittner et al. (2014))

Fig. 6.37 Industrial installation of ST separator (Adapted from Bittner et al. (2014))

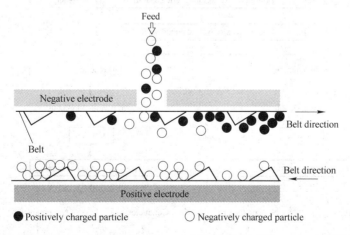

Fig. 6.38 Separation zone of ST separator (Adapted from Bittner et al. (2014))

power plants (Bittner et al. ,2014). On a pilot plant scale (3~6t/h), it has also been shown to be effective at beneficiating industrial minerals such as separating quartz from calcite (89% recovery, 99% grade) and magnesite from talc (77% recovery, 95% grade) (Bittner et al. ,2014).

6.6.4 Example Flowsheets

Earlier in this chapter the possibility of combined magnetic and electrical separation was noted, particularly in the processing of heavy mineral sand deposits. Table 6.3 shows some of the common minerals present in such alluvial deposits, along with their properties, related to magnetic and electrical separation. Mineral sands are commonly mined by floating dredges, feeding floating concentrators at up to 2000t/h. Such concentrators, consisting of a complex circuit of sluices, spirals, or Reichert cones, upgrade the heavy mineral content to around 90%, the feed grades varying from less than 2%, up to 20% heavy mineral in some cases. The gravity pre-concentrate is then transferred to the separation plant for recovery of heavy minerals by a combination of gravity, magnetic, and electrical (typically HTR) separation.

Table 6.3 Magnetic and Electrical Behavior of Typical Heavy Mineral Sands Components

Magnetics	Magnetite—C	Ilmenite—C	Garnet—NC	Monazite—NC
Nonmagnetics	Rutile—C	Zircon—NC	Quartz—NC	

Note: C—conductor; NC—nonconductor.

Mineral sands flowsheets vary according to the properties of the minerals present, wet magnetic separation often preceding high-tension separation where magnetic ilmenite is the dominant mineral, for example. A generalized flowsheet is shown in Fig. 6.39. Low-intensity drum separators

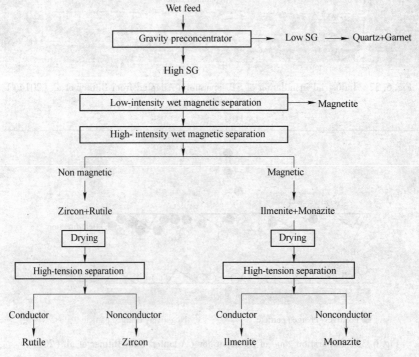

Fig. 6.39 Typical heavy mineral sand flowsheet

remove any magnetite from the feed, after which high-intensity wet magnetic separators separate the monazite and ilmenite from the zircon and rutile. Drying of these two fractions is followed by HTR separation to produce final products, although further cleaning is sometimes carried out by ESP separators. For example, screen electrostatic separators (Fig. 6.34(b)) may be used to clean the zircon and monazite concentrates, removing fine conducting particles from these fractions. Similarly, plate electrostatic separators (Fig. 6.34(a)) could be used to reject coarse nonconducting particles from the rutile and ilmenite concentrates.

Fig. 6.40 shows a simplified circuit used to process heavy minerals, on the west coast of Australia (Benson et al., 2001).

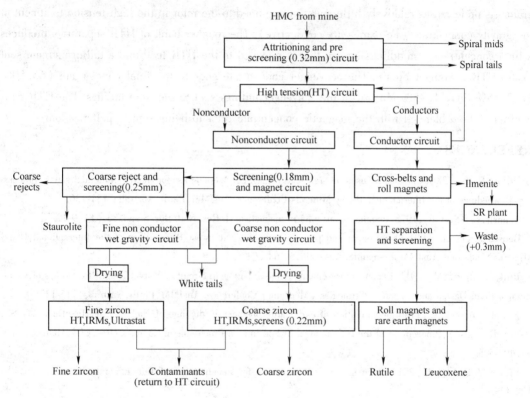

Fig. 6.40 Simplified mineral sands circuit used by Tronox Limited (formerly Tiwest Joint Venture) at the Chandala processing plant, Western Australia (Adapted from Benson et al. (2001))

The heavy mineral concentrate is first divided into conductive and nonconductive streams using HTR separators. The conductors are treated using cross-belt and roll magnetic separators to remove the ilmenite as a magnetic product. The nonmagnetic stream is cleaned with highintensity roll and rare earth magnets to separate the weakly paramagnetic leucoxene from diamagnetic rutile. The nonconductors undergo another stage of wet gravity separation to remove quartz and other low-specific-gravity contaminants, before sizing and cleaning using HTR, ESP and Ultrastat separators to produce fine and coarse zircon products. Similar flowsheets are used in South-East Asia for the treatment of alluvial cassiterite deposits, which are also sources of minerals such as ilmenite, monazite, and zircon.

In the case of the Cliffs-Wabush mine (discussed in Section 6.5.2), the gravity concentrate from the spirals bank is cleaned by a series of HTR separators (Damjanovic and Goode, 2000). The spiral concentrate is first filtered and dried before being fed to 54 primary Carpco HTR separators (with a total of 288 rotors), with the tailings from the rougher HTR separators fed to six scavenger HTR separators (total of 24 rotors). Each rotor is 10ft long with a roll diameter of 14in. and is operated at a rotor speed of 100r/min, electrode voltage of 23-25kV, and a feed rate of 2.54t/h. Through this separation, the nonconductive quartz gangue is pinned to the roll, with the valuable iron oxide mineral thrown from the roll. Cleaning of the gravity concentrate with HTR separators is preferred as relatively little material is pinned to the rotor in the high-tension treatment of the gravity concentrate (Fe-oxides are conductive). The rougher bank of HTR separators produces a final concentrate, a middlings stream that is recycled to the HTR feed and a tailings stream sent to the HTR scavenger circuit. The scavenger concentrate goes to the final concentrate (65.50% Fe, 2.55% SiO_2, 1.95% Mn), with the scavenger tailings sent to the final tailings. The HTR concentrate is then blended with the magnetic concentrate prior to being sent to pelletization.

REFERENCES

Arvidson, B. R., 2001. The many uses of rare-earth magnetic separators for heavy minerals sands processing. Proceedings of the International Heavy Minerals Conference. AusIMM, Perth, Australia, 131-136.

Bennett, L. H., et al., 1978. Comments on units in magnetism. J. Res. Natl. Bur. Stand. 83 (1), 9-12.

Benson, S., et al., 2001. Quantitative and process mineralogy at tiwest. Proceedings of the International Heavy Minerals Confernce. AusIMM, Fremantle, Australia, 59-68.

Bird, A., Briggs, M., 2011. Recent improvements to the gravity gold circuit at Marvel Loch. Proceedings of Metallurgical Plant Design and Operating Strategies (MetPlant) Conference. AusIMM, Perth, Australia, 115-137.

Biss, R., Ayotte, N., 1989. Beneficiation of carbonatite ore-bearing niobium at Niobec Mine. In: Dobby, G. S., Rao, S. R. (Eds.), Proceedings of the International Symposium on Processing of Complex Ores. CIM, Halifax, Canada, 497-506.

Bittner, J. D., et al., 2014. Triboelectric belt separator for beneficiation of fine minerals. Procedia Eng. 83, 122-129.

Butcher, D. A., Rowson, N. A., 1995. Electrostatic separation of pyrite from coal. Mag. Elect. Sep. 9, 19-30.

Corrans, I. J., 1984. The performance of an industrial wet high-intensity magnetic separator for the recovery of gold and uranium. J. S. Afr. Inst. Min. Metall. 84 (3), 57-63.

Corrans, I. J., Svoboda, J., 1985. Magnetic separation in South Africa. Mag. Sep. News. 1, 205-232.

Damjanovic, B., Goode, J. R., 2000. Canadian Milling Practice, Special Vol. 49. CIM, Montre'al, QC, Canada.

Dance, A. D., Morrison, R. D., 1992. Quantifying a black art: the electrostatic separation of mineral sands. Miner. Eng. 5 (7), 751-765.

Dascalescu, L., et al., 1993. Corona-electrostatic separation: an efficient technique for the recovery of metals and plastics from industrial wastes. Mag. Elect. Sep. 4, 241-255.

Davis, E. W., 1921. Magnetic Concentrator. US Patent No. 1474624.

Dobbins, M., Sherrell, I., 2010. Significant developments in dry rare-earth magnetic separation. Mining Eng. 62 (1), 49-54.

Dobbins, M. , et al. , 2009. Recent advances in magnetic separator designs and applications. Proceedings of the Seventh International Heavy Minerals Conference. South African Institute of Mining and Metallurgy, Johannesburg, South Africa, 63-70.

Dwari, R. K. , et al. , 2009. Characterisation of particle tribo-charging and electron transfer with reference to electrostatic dry coal cleaning. Int. J. Miner. Process. 91, 100-110.

Edward, D. , et al. , 1995. The motion of mineral sand particles on the roll in high tension separators. Mag. Elect. Sep. 6, 69-85.

Elder, J. , Yan, E. , 2003. EForce—Newest generation of electrostatic separator for the mineral sands industry. Proceedings of Heavy Minerals 2003 Conference. South African Institute of Mining and Metallurgy, Johannesburg, South Africa, 63-70.

Eriez, 2003. Rare earth roll magnetic separators. Eriez Mag. 1-8.

Eriez, 2008. WHIMS Separators. Eriez Manufacturing Company, Erie, PA, USA, 1-8.

Eriez, Gzrinm, 2014. Reliable WHIMS with Maximum Recovery. Eriez Manufacturing Company, Erie, PA, USA, 1-12.

Germain, M. , et al. , 2003. The application of new design concepts in high tension electrostatic separation to the processing of mineral sands concentrates. Proceedings of the International Heavy Minerals Conference. South African Institute of Mining and Metallurgy, Johannesburg, South Africa, 101-106.

Gillet, G. , Diot, F. , 1999. Technology of superconducting magnetic separation in mineral and environmental processing. Miner. Metall. Process. 16 (3), 1-7.

Gillet, G. , et al. , 1999. Removal of heavy metal ions by superconducting magnetic separation. Sep. Sci. Technol. 34 (10), 2023-2037.

Jiles, D. , 1990. Introduction to Magnetism and Magnetic Materials. Chapman & Hall, London, UK.

Karapinar, N. , 2003. Magnetic separation of ferrihydrite from waste water by magnetic seeding and high-gradient magnetic separation. Int. J. Miner. Process. 71, 45-54.

Kelly, E. G. , Spottiswood, D. J. , 1989a. The theory of electrostatic separations: a review—part II. Particle charging. Miner. Eng. 2 (2), 193-205.

Kelly, E. G. , Spottiswood, D. J. , 1989b. The theory of electrostatic separations: a review—part III. The separation of particles. Miner. Eng. 2(3), 337-349.

Kelly, E. G. , Spottiswood, D. J. , 1989c. The theory of electrostatic separations: a review—part I. Fundamentals. Miner. Eng. 2 (1), 33-46. Kopp, J. , 1991. Superconducting magnetic separators. Mag. Elect. Sep. 3, 17-32.

Lawver, J. E. , Hopstock, D. M. , 1974. Wet magnetic separation of weakly magnetic minerals. Miner. Sci. Eng. 6 (3), 154-172.

Li, T. X. , et al. , 1999. Dry triboelectrostatic separation of mineral particles: a potential application in space exploration. J. Electrostat. 47, 133-142.

Lockhart, N. C. , 1984. Dry beneficiation of coal. Powder Technol. 40, 17-42.

Malati, M. A. , 1990. Ceramic superconductors. Mining Mag. 163 (Dec.), 427-431.

Manouchehri, H. R. , 2010. Triboelectric charge and separation characteristics of industrial minerals. Proceedings of the 25th International Mineral Processing Congress (IMPC), Brisbane, Australia, 1009-1021.

Manouchehri, H. R. , Fawell, S. , 2002. Electrophysical properties, tribo-electric charge characteristics and separation behaviour of calcite, quartz and wollastonite minerals. Proceedings of the SME Annual Meeting. SME, Phoenix, AZ, USA, 1-10.

Manouchehri, H. R. , et al. , 1999. Changing potential for the electrical beneficiation of minerals by chemical pretreatment. Miner. Metall. Process. 16 (3), 14-22.

Manouchehri, H. R., et al., 2000. Review of electrical separation methods. Part 1: fundamental aspects. Miner. Metall. Process. 17 (1), 23-36.

McAndrew, J., 1957. Calibration of a Frantz isodynamic separator and its application to mineral separation. Proc. Aust. Inst. Min. Met. 181, 59-73.

Metso, 2014a. Magnetic Separators for Dense Media Recovery. Metso Minerals, Sweden, 1-8.

Metso, 2014b. Drum Separators. Metso Minerals, Sweden, 1-4.

Metso, 2014c. High Gradient Magnetic Separators—HGMS Cyclic. Metso Minerals, Sweden, 1-5.

Metso, 2014d. High Gradient Magnetic Separators—HGMS Continuous. Metso Minerals, Sweden, 10-4.

Murariu, V., Svoboda, J., 2003. The applicability of Davis tube tests to ore separation by drum magnetic separators. Phys. Sep. Sci. Eng. 12 (1), 1-11.

Norrgran, D. A., Marin, J. A., 1994. Rare earth permanent magnet separators and their applications in mineral processing. Miner. Metall. Process. 11 (1), 41-45.

Novotny, D., 2014. Technological advances in magnetic separation-Presented as part of the CMP 2014 "Iron Ore Processing in Canada" short course. The 46th Annual Meeting of the Canadian Mineral Processors Conference, Ottawa, ON, Canada.

Oberteuffer, J., 1974. Magnetic separation: a review of principles, devices, and applications. IEEE Trans. Magn. 10 (2), 223-238.

Outotec, 2013. SLon vertically pulsating high-gradient magnetic separator. 1-4. Outotec., www.outotec.com/.

Rayner, J. G., Napier-Munn, T. J., 2000. The mechanism of magnetics capture in the wet drum magnetic separator. Miner. Eng. 13 (3), 277-285.

Shao, Y., et al., 1996. Wet high intensity magnetic separation of iron minerals. Magn. Elect. Sep. 8 (1), 41-51.

Singh, V., et al., 2013. Particle flow modeling of dry induced roll magnetic separator. Powder Technol. 244, 85-92.

Song, S., et al., 2002. Magnetic separation of hematite and limonite fines as hydrophobic flocs from iron ores. Miner. Eng. 15, 415-422.

Svoboda, J., 1987. Magnetic Methods for the Treatment of Minerals. Elsevier, Amsterdam, North Holland, Netherlands.

Svoboda, J., 1993. Separation of particles in the corona-discharge field. Magn. Elect. Sep. 4, 173-192.

Svoboda, J., 1994. The effect of magnetic field strength on the efficiency of magnetic separation. Miner. Eng. 7 (5-6), 747-757.

Svoboda, J., Fujita, T., 2003. Recent developments in magnetic methods of material separation. Miner. Eng. 16 (9), 785-792.

Tawil, M. M. E., Morales, M. M., 1985. Application of wet high intensity magnetic separation to sulphide mineral beneficiation. In: Zunkel, A. D. (Ed.), Complex Sulfides: Processing of Ores, Concentrates and By-products. SEM, Pennsylvania, PA, USA, 507-524.

Todd, I. A., Finch, J. A., 1984. Measurement of susceptibility using a wet modification to the Frantz Isodynamic Separator. Can. Met. Quart. 23 (4), 475-477.

Unkelbach, K.-H., Kellerwessel, H., 1985. A superconductive drum type magnetic separator for the beneficiation of ores and minerals. Congre's international de mine'ralurgie, Cannes, France, 371-380.

Wasmuth, H.-D., Unkelbach, K.-H., 1991. Recent developments in magnetic separation of feebly magnetic minerals. Miner. Eng. 4 (7-11), 825-837.

Watson, J. H. P., 1994. Status of superconducting magnetic separation in the minerals industry. Miner. Eng. 7 (5-6), 737-746.

Watson, J. H. P., Beharrell, P. A., 2006. Extracting values from mine dumps and tailings. Miner. Eng. 19, 1580-1587.

White, L., 1978. Swedish symposium offers iron ore industry an over-view of ore dressing developments. Eng. Min. J. 179 (4), 71-77.

Xiong, D. -H., 1994. New development of the SLon vertical ring and pulsation HGMS separator. Magn. Elect. Sep. 5, 211-222.

Xiong, D. -H., 2004. SLon magnetic separators applied in the ilmenite processing industry. Phys. Sep. Sci. Eng. 13 (3-4), 119-126.

Chapter 7   Dewatering

7.1   INTRODUCTION

With few exceptions, most mineral-separation processes involve the use of substantial quantities of water and the final concentrate has to be separated from a pulp in which the water solids ratio may be high. Dewatering, or solid-liquid separation, produces a relatively dry concentrate for shipment. Partial dewatering is also performed at various stages in the concentrator, so as to prepare the feed for subsequent processes. Sometimes tailings are also dewatered.

Dewatering methods can be broadly classified into three groups: sedimentation (gravity and centrifugal), filtration, and thermal drying. Dewatering in mineral processing is normally a combination of these methods, an example being Fig. 7.1. The bulk of the water is first removed by thickening, which produces a thickened pulp of perhaps 55%~65% solids by weight. Up to 80% of the water can be separated at this stage. Filtration of the thickened pulp then produces a moist filter cake of between 80% and 90% solids, which may require thermal drying to produce a final product of about

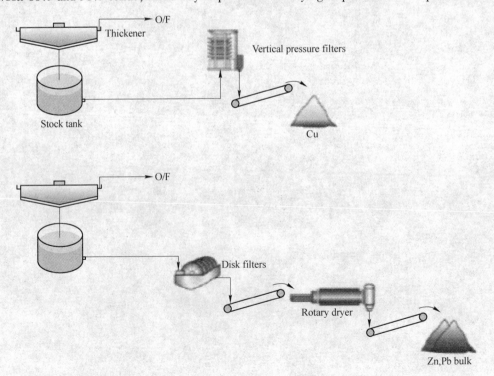

Fig. 7.1   Example of thickening, filtering, and drying (Brunswick Mine concentrator)

95% solids by weight.

The principles of solid-liquid separation, testing, equipment sizing, and operation are covered in detail by Con cha (2014) and Mular et al. (2002); innovations are discussed by Mc Caslin et al. (2014), and specifically in coal preparation (washing) plants by Luttrell (2014).

7.2 GRAVITATIONAL SEDIMENTATION

Gravity sedimentation or thickening is the most widely applied dewatering technique in mineral processing, and it is a relatively cheap, high-capacity process. The thickener is used to increase the concentration of the suspension by sedimentation, accompanied by the formation of a clear liquid (supernatant). The principal type of thickener consists of a cylindrical, largely open tank from which the clear liquid is taken off at the top (thickener "overflow"), and the suspension is transported by rotating rakes to discharge at the bottom ("underflow") (Schoenbrunn and Laros, 2002). These are conventional thickeners and their variants, and they form the main content of this section of the chapter. Other sedimentation devices, including centrifuges, are briefly covered at the end of the section.

7.2.1 Sedimentation of Particles

Sedimentation is most efficient when there is a large density difference between liquid and solid. This is always the case in mineral processing where the carrier liquid is water. Sedimentation cannot always be applied in hydrometallurgical processes, however, because in some cases the carrier liquid may be a high-grade leach liquor having a density approaching that of the solids.

Settling of solid particles in a liquid produces a clarified liquid which can be decanted, leaving a thickened slurry. The process is illustrated in Fig. 7.2 using the common laboratory test, that is, batch sedimentation in a graduated cylinder (cylinder test or jar test).

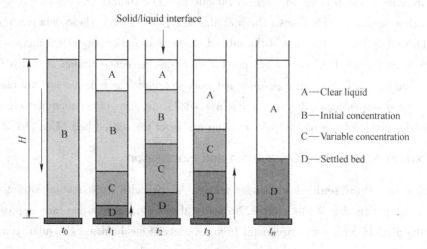

Fig. 7.2 Batch settling process (t_i is time)

The settling rates of particles in a fluid are governed by Stokes' or Newton's laws (Chapter 3). Factors that affect sedimentation include: particles size and shape, weight and volume content of solids, fluid viscosity, and specific gravity of solids and liquid. Very fine particles, of only a few micrometers diameter, settle extremely slowly by gravity alone, and centrifugal sedimentation may have to be performed. Alternatively, the particles may be aggregated or flocculated, into relatively large clumps, called flocs that settle out more rapidly (Section 7.2.2).

Batch settling tests are undertaken to size thickeners, which operate continuously, and to test performance of operating thickeners. The data for analysis are presented as solid-liquid interface height versus time (Fig. 7.3).

The start of the settling curve represents the interface (or mud-line) between zone A, clear water (clarified supernatant), and zone B at the initial concentration; the settling rate is nearly constant. The critical sedimentation point indicates the loss of zone B and the interface now corresponds to that between zone A and the variable concentration zone C. The settling rate now decreases as the

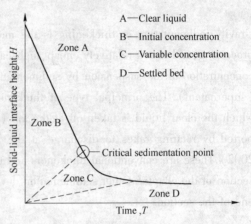

Fig. 7.3 Example batch settling curve
(Interface or mud-line vs. time)

concentration in zone C increases, hindering the settling process. When zone C is lost, the settling rate approaches zero, the interface only slowly decreasing as water is squeezed out by the weight of the particles compressing the bed. The settling curve should be independent of vessel geometry, avoiding, for example, wall effects by using a too small diameter cylinder. Concha (2014) describes a setup to measure settling characteristics in five cylinders simultaneously (Sedi Rack).

Additional detail on suspension settling behavior can be obtained by measuring local solids concentration in zones C and D using, for example, mobile radiation sources and detectors (X-rays, gamma-rays) with appropriate calibrations (Owen et al.,2002;Kurt,2006). These data can reveal channeling in the settling bed. In the case of flocculated suspensions, other parameters that can be determined include compressibility and the gel point, defined as the concentration at which flocs come into contact and start to form a self-supporting net work structure. These parameters are used in theories of flocculated suspensions (Bus call and White,1987). An array of conductivity sensors along a settling cylinder provides data on the concentration profile of the settled bed (Con cha,2014).

7.2.2 Particle Aggregation: Coagulation and Flocculation

Fine particles, say-10μm, settle slowly under gravity. All particles exert mutual attractive forces, known as London-Van der Waals' forces. Normally these attractive forces are opposed by the charge on the particle surface that originates from a variety of mechanisms. Coagulation and flocculation refer to processes that cause particles to aggregate or agglomerate (i.e., adhere or cluster together) to increase settling rates. The cluster is variously called an aggregate, agglomerate, or more commonly a floc. Often used synonymously, coagulation is associated with modification of particle surface charge to cause aggregation, while flocculation involves addition of long-chain polymers that bind ("bridge") particles together. This difference leads to two classes of aggregating reagents: coagulants and flocculants.

7.2.2.1 Surface Charge

Particles in water always exhibit a surface charge. In a given system the electrical charge on the particle surfaces will be of the same sign and this causes mutual repulsion, which slows settling by keeping particles apart and inconstant motion. In mineral processing most aqueous systems are alkaline and the charge is usually taken to be negative; a negative charge will be assumed in the discussion that follows. The charge at the particle surface affects the distribution of solute ions nearby (Fig. 7.4(a)). The solute ions form two layers: an inner layer that comprises an excess of cations (assuming the surface is negatively charged) more or less bound to the surface, the bound layer (or Stern layer); and a second layer that contains more loosely attracted ions, the diffuse layer. This depiction is referred to as the electrical double layer model. The charge is normally expressed as a potential. Fig. 7.4(a) illustrates the distribution of ions (upper) and the corresponding electrical potential versus distance from the particle surface (lower).

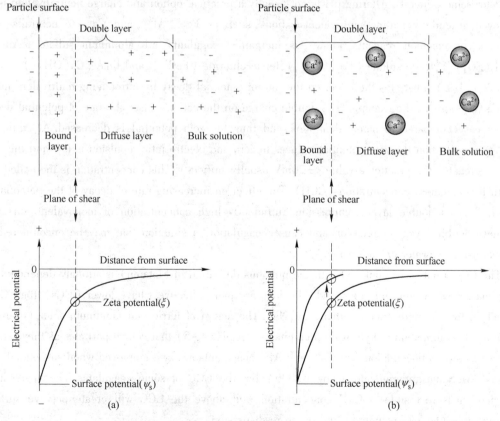

Fig. 7.4 Distribution of ions at a charged surface (upper) and
potential as a function of distance from negatively charged surface (lower):
(a) dilute solution of monovalent ions, and (b) after adding Ca^{2+} ions (as lime)

The potential decays with distance from the surface (surface potential) to reach zero at the bulk solution. Since suspension behavior involves motion of the particlerel ative to the water, a layer of water is associated with the moving particle. The relevant potential then is not the surface potential, but that at the plane of shear (the boundary between the water layer carried by the particle and the bulk water), termed the zeta potential. The plane of shear is sometimes taken to coincide with the boundary between the Stern and diffuse layers. The zeta potential is measured by the common methods that track the motion of the particle under the influence of an electrical field (electrophoresis).

Zeta potential values range up to about 200mV. A zeta potential of 20mV or more (negative or positive) will tend to keep fine particles dispersed by electrostatic repulsion. There are two ways the charge can be manipulated to cause aggregation: reducing the zeta potential to near zero to remove the repulsive force (charge neutralization); or creating conditions to cause electrostatic attraction (electrostatic coagulation).

7.2.2.2 Charge Neutralization

In Flotaion we learned that zeta potential can be reduced to zero by adjusting pH to the iso-electric point (IEP). In fact, determining the pH giving maximum settling rate can be used to identify the IEP for some minerals. Altering pH is not usually a practical option and charge neutralization is achieved instead by adding multivalent cations, such as Fe^{3+}, Al^{3+}, and Ca^{2+} to neutralize the charge. The common reagents, known as inorganic coagulants, are aluminum sulfate (Alum) ($Al_2(SO_4)_3$), ferric sulfate ($Fe_2(SO_4)_3$), ferric chloride ($FeCl_3$), and lime ($Ca(OH)_2$).

Fig. 7.4(b) illustrates the effect on the rate of potential decay by introducing multivalent ions, using Ca^{2+} ions as the example. The double charge on the cation causes the rate of potential decay to increase compared to monovalent ions, and thus the zeta potential is decreased. At sufficient Ca^{2+} concentration the zeta potential reduces to zero and electrostatic repulsion is lost and the system aggregates (in practice a value $<\pm 20mV$ usually suffices). This concentration is the called the critical coagulation concentration (CCC). The effect on increasing rate of decay of the potential is referred to as double layer compression. Sufficiently high concentration of monovalent ions can achieve double layer compression and cause coagulation, a situation that may be encountered in plants using seawater or bore water.

The CCC (not to be confused with the previous definition in Flotaion) is strongly dependent on the charge on the ion, thus Fe^{3+} and Al^{3+} ions are more effective (have lower CCCs) than Ca^{2+} which in turn is more effective than, say, Na^+. The action of ferric and aluminum salts, however, also involves formation of hydroxide precipitates (at pH>4~5) that collect particles as they settle, known as sweep flocculation. The Fe^{3+} and Al^{3+} coagulants are best employed when suspended solids is low. Somasundaran and Wang (2006) list the CCC of some coagulants. The aggregation mechanism is reversible; a Ca^{2+} concentration well above the CCC will create positive surface charge (positive zeta potential) and cause re-dispersion.

7.2.2.3 Electrostatic Coagulation

Coagulation by electrostatic attraction occurs when two particle types have opposite charge and there is mutual attraction. Talc presents an example of a single mineral subject to electrostatic coagulation, as the charge on the basal plane (face) is usually negative and the charge on the edge (at least below about pH 10) is positive and aggregation by a face-edge ("house-of-cards") arrangement occurs. Electrostatic coagulation in the present case is achieved by adding organic coagulants. These are polymers of low molecular weight (3,000 to 1million) with cationic functional groups. Adsorption creates regions of positive charge ("patches") that promote electrostatic attraction with the polymer-free negative regions on other particles, as depicted in Fig. 7.5. Also known as patch flocculation, like inorganic coagulants, over-dosing can cause redispersion, as the patches spread to cover the surface and the charge becomes uniformly positive.

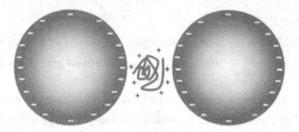

Fig. 7.5 Coagulation through electrostatic attraction (electrostatic coagulation)

Three families of organic coagulants are primarily used: polyamines, poly DADMAC, and dicyandiamide. More complete lists are given by Somasundaran and Wang (2006) and Bulatovic (2007).

7.2.2.4 Flocculation

Creating a more open structure than coagulation, flocculation involves the use of long-chain organic polymers of high molecular weight (> 1million) to form molecular bridges between particles (Hogg, 2000; Pearse, 2003; Tripathy and De, 2006; Usher et al. , 2009). Formerly natural products such as starch derivatives and polysaccharides, they are now increasingly synthetic materials, loosely termed polyelectrolytes. Bridging is illustrated in Fig. 7.6(a) for a cationic polymer. In practice, many such inter-particle bridges are formed, linking a number of particles together. Fig. 7.6(b) shows the co-use of coagulant (Ca^{2+}) and an anionic polymer. The length of a completely uncoiled polymer is about 1μm, up to perhaps a few tens of μm for the longest chains. Bridging with other particles stops when the floc size reaches about 10mm (Rushton et al. , 2000).

It would be expected that, since most suspensions encountered in the minerals industry contain negatively charged particles, cationic polyelectrolytes would be most suitable. Although this gives some level of charge neutralization, and aids attraction of the polymer to the particle surface, it is not necessarily true for the bridging role of the flocculant. For bridging, the polymer must be strongly attached, and this is promoted by chemical adsorption (formation of chemical bonds) through, for example, amide groups (e. g. , $CONH_2$). In other words, the charge on the polymer is less important and the majority of commercially available polyelectrolytes are anionic, since these tend to be of higher molecular weight than the cationics, and are less expensive.

Bridging requires that the polymer be strongly bonded to one particle, but have other bonding

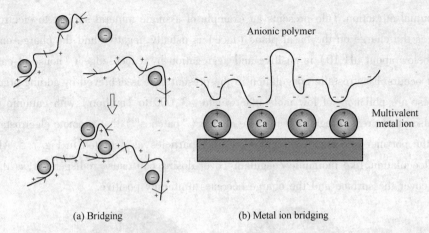

Fig. 7.6 Flocculation mechanisms using organic polymers: (a) bridging by high MW cationic polymer, and (b) co-use of Ca^{2+} adsorption and bridging by anionic polymer

sites available for other particles. Excess polymer tends to adsorb on one particle and this can promote redispersion. The optimum polymer concentration and pH requires laboratory testing. Physical conditions, such as agitation and pumping, can also cause redispersion by breaking down the flocs. This is one reason why flocculants are generally not used with hydrocyclones and have limited application in centrifugal dewatering. In comparison, aggregates produced by coagulation will reform after disruption.

The polyacrylamides (abbreviated as PAM) are the most common flocculants. They can be manufactured to have non-ionic, anionic, or cationic character. Polyacrylamide formed from the acrylamide monomer ($CH_2CHCONH_2$) is non-ionic (Fig. 7.7(a)). (Being made of only one monomer type it is a homopolymer.) An anionic polymer can be made by hydrolysis of the non-ionic PAM, but more usually by co-polymerizing acrylamide with acrylic acid ($CH_2 = CHCOOH$) monomer (i.e., making a co-polymer), where some of the side chains lose a proton to become negatively charged (Fig. 7.7(b)). A cationic polymer is made by co-polymerizing with a cationic monomer.

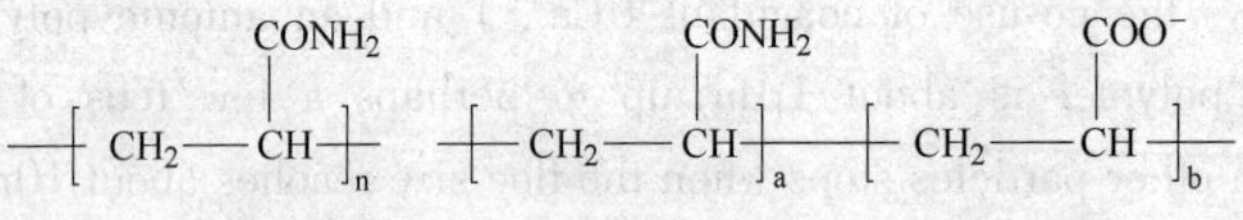

$$\left[CH_2 - CH \right]_n \quad \left[CH_2 - CH \right]_a \quad \left[CH_2 - CH \right]_b$$
$$\quad\quad | \quad\quad\quad\quad\quad\quad | \quad\quad\quad\quad\quad\quad |$$
$$\quad CONH_2 \quad\quad\quad\quad CONH_2 \quad\quad\quad\quad COO^-$$

(a)   (b)

Fig. 7.7 Polyacrylamide flocculants:
(a) uncharged polymer comprising n acrylamide monomers, and
(b) an anionic polymer made by co-polymerization with acrylic acid

The manufacture gives a certain degree of ionic character or charge density, which refers to the percentage of the monomer segments that carry a charge. It is now possible to obtain water soluble PAM products with a wide range of ionic character, varying from 100% cationic content through non-ionic to 100% anionic content a wide range of ionic character, varying from 100% cationic content through non-ionic to 100% anionic content and with molecular weights from several thousand to over 10 million, the highest among the synthesized polymers. Flocculants other than PAM (e. g. , polyethylene-imines, polyamides-amines) are used under special conditions.

Although the addition of flocculants can lead to significant improvements in sedimentation rate, flocculation is generally detrimental to final consolidation of the sediment. Large flocs promote settling and are desirable for clarification and thickening. Floc density is of secondary importance in these processes. Conversely, dense flocs are most appropriate for consolidation of the sediment, and size is of lesser importance in this stage. Therefore the optimization of solid-liquid separation processes requires careful control of floc size and structure. If thickening is followed by filtration, the choice of flocculant may be important. Flocculants are widely used as filter aids. However, the specific requirements of a flocculant used to promote sedimentation are not necessarily the same as for one used as a filter aid; for example, flocs formed with high molecular weight products are relatively large, trapping water within the structure and increasing the final moisture content of the filter cake.

Laboratory batch cylinder tests are commonly used to assess the effectiveness of flocculants. Reproducibility of such tests is often poor and there is almost always conflict in determining an optimum dosage of flocculant between full-scale operations and laboratory tests (Scales et al. , 2015). Methods of improving reproducibility and better approximating conditions in the thickener, such as shear rate, continue to be developed (Farrow and Swift, 1996; Scales et al. , 2015; Parsapour et al. , 2014).

In practice, polyelectrolytes are normally made up of stock solutions of about 0.5% ~ 1%, which are diluted to about 0.1% (maximum) before adding to the slurry. The diluted solution must be added at enough points in the stream to ensure its contact with every portion of the system. Agitation is essential at the addition points, and shortly thereafter, to assist in flocculant dispersion in the process stream. Care should be taken to avoid severe agitation after the flocs have been formed. The age of the stock solution can have a significant effect on flocculant performance (Owen et al. , 2002).

7.2.3 Thickener Types

There are four types of thickener that have a more or less open tank design (Concha, 2014), shown in Fig. 7.8: conventional (a), high rate (b), high density (c), and paste (d) thickeners. High rate (or high capacity) is the term applied to thickeners processing very high throughput by optimization of flocculation. In that regard, it may be more appropriate to talk of different thickener operation rather than thickener type. High density and paste thickeners are similar to conventional thickeners, but with steeper cone angles and higher sided tanks. The extra height increases the pressure on the sediment bed and thus gives higher density underflow. Applications of paste thickeners include mine backfill and tailings disposal. The clarifier is similar in design, but is less robust, handling suspensions of much lower solid content than the thickener and designed for removal of solids rather than their compaction (Seifert and Bowersox, 1990). Given the basic similarity in design, just the conventional thickener and its variants will be described in any detail.

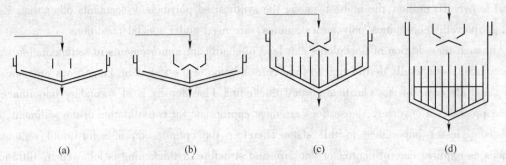

Fig. 7.8 Types of thickener equipment (Deep Cones is a registered trademark of FLSmidth)
(a) conventional; (b) high rate; (c) high density; (d) deep Cone® paste

The conventional thickener consists of a cylindrical ank, the diameter ranging from about 2 to 200m in diameter, ad of depth 1~7m. The clarified liquid overflows aperipheral launder, while the solids, which settle over the ntire bottom of the tank, are withdrawn as a thickened pulp from an outlet at the center. The zones in the thickener Fig. 7.9 mirror those recognized in the batch cylinder sedimentation test (Fig. 7.2).

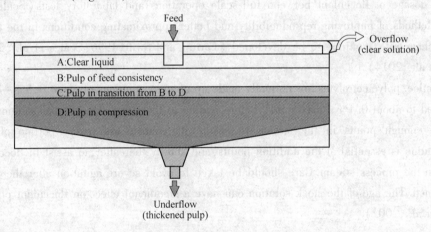

Fig. 7.9 Concentration zones in a thickener

Pulp is fed into the center via a feedwell placed up to 1m below the surface. The feedwell is a small concentric cylinder with several key functions, including (Loan et al., 2009; Owen et al., 2009; Lake and Summerhays, 2012): controlling momentum dissipation, deaerating feed slurry, diluting feed (if required), optimizing flocculation, and ensuring even distribution of the feed stream into the thickener. Within the tank are one or more rotating radial arms, from each of which are suspended a series of blades, shaped so as to rake the settled solids toward the central outlet. On most thickeners today these arms rise automatically if the torque exceeds a certain value, thus preventing damage due to overloading. The blades also assist the compaction of the settled particles and produce a thicker underflow than can be achieved by simple settling by assisting the removal of water. In paste thickeners the high yield stress of the suspension can lead to a phenomenon known as "rotating beds," "doughnuts," or "islands", which should be avoided (Arbuthnot et al.,

2005). This is associated with the presence of large aggregates that form ahead of and are compacted by the rotating blades. The solids in the thickener move continuously downwards, and then inwards toward the thickened underflow outlet, while the liquid moves upwards and radially outwards. In general, there is no region of constant composition in the thickener.

Thickener tanks are constructed of steel, concrete, or a combination of both, steel being most economical in sizes of less than 25m in diameter. The tank bottom is often flat (e. g. ,Fig. 7. 10), while the mechanism arms are sloped toward the central discharge. With this design, settled solids must "bed-in" to form a false sloping floor. Steel floors are rarely sloped to conform to the rake arms because of expense. Concrete bases and sides become more common in the larger sized tanks. In many cases the settled solids, because of particle size, tend to slump and will not form a false bottom. In these cases the floor should be concrete and poured to match the slope of the arms. Tanks may also be constructed with sloping concrete floors and steel sides. Earth bottom thickeners are also in use, which are generally considered to be the lowest cost solution for thickener bottom construction (Hsia and Reinmiller,1977).

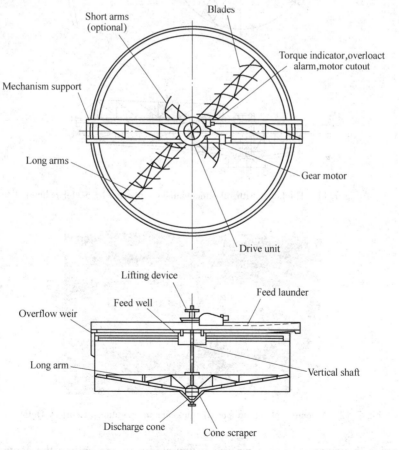

Fig. 7. 10   Thickener with mechanism supported by superstructure

The method of supporting the raking mechanism depends primarily on the tank diameter. In thickeners of diameter less than about 45m, the drive head is usually supported on a superstructure spanning the tank, with the arms being attached to the drive shaft. Such machines are referred to as bridge or beam thickeners (Fig. 7.10). The underflow is usually drawn from the apex of a cone located at the center of the sloping bottom. A common arrangement for larger thickeners is to support the drive mechanism on a stationary steel or concrete center column. In most cases, the rake arms are attached to a drive cage, surrounding the central column, which is connected to the drive mechanism. The thickened solids are discharged through an annular trench encircling the center column (Fig. 7.11). Fig. 7.12 shows a thickener of this type in operation.

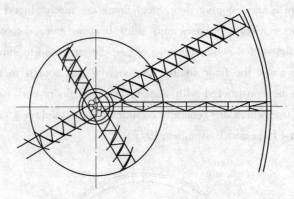

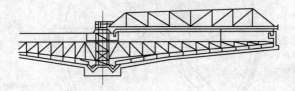

Fig. 7.11 Thickener with rake mechanism supported by center column

Fig. 7.12 A center-column supported thickener in operation (Courtesy Outotec)

In the traction thickener, a single long arm is mounted with one end on the central support column, while to the other end are fixed traction wheels that run on a rail on top of the tank wall. The wheels are driven by motors that are mounted on the end of the arm and which therefore travel around with it. This is an efficient and economical design since the torque is transmitted through a long lever arm by a simple drive. They are manufactured in sizes ranging up to 200m in diameter.

Cable thickeners have a hinged rake arm fastened to the bottom of the drive cage or center shaft. The rake arm is pulled by the cables connected to a torque or drive arm structure, which is rigidly connected to the center shaft at a point just below the liquid level. The hinge allows the rake to automatically lift when torque rises, which enables the rake arm to find its own efficient working level in the sludge, where the torque balances the rake weight. A feature of the design is the relatively small surface area of the raking mechanism.

In all thickeners, the speed of the raking mechanism is normally about 8m/min at the perimeter, which corresponds to about 10r/h for a 15m diameter thickener. Energy consumption is thus extremely low, such that even a 60m unit may require only a 10kW motor. Wear and maintenance costs are correspondingly low.

The underflow is usually withdrawn by pumping, although in clarifiers the material may be discharged under the hydrostatic head in the tank. The underflow is usually collected in a sludge-well in the center of the tank bottom, from where it is removed via piping through an underflow tunnel. The underflow lines should be as short and as straight as possible to reduce the risk of choking, and this can be achieved, with large tanks, by taking them up from the sludge-well through the center column to pumps placed on top, or by placing the pumps in the base of the column and pumping up from the bottom. This has the advantage of dispensing with the expensive underflow tunnel. A development of this is the caisson thickener, in which the rake assembly is supported on hydrostatic bearings and the center column is enlarged sufficiently to house a central control room; the pumps are located in the bottom of the column, which also contains the mechanism drive heads, motors, control panel, underflow suction, and discharge lines. The caisson concept has lifted the possible ceiling on thickener sizes.

Underflow pumps are often of the diaphragm type. These are positive action pumps for medium heads and volumes, and are suited to the handling of thick viscous fluids. They can be driven by an electric motor through a crank mechanism, or by directly acting compressed air. A flexible diaphragm is oscillated to provide suction and discharge through non-return valves, and variable speed can be achieved by changing either the oscillating frequency or the stroke. In some plants, variable-speed pumps are connected to nucleonic density gauges on the thickener underflow lines, which control the rate of pumping to maintain a constant underflow density. The thickened underflow is commonly pumped to filters for further dewatering.

Thickeners often incorporate substantial storage capacity so that, for instance, if the filtration section is shut down for maintenance, the concentrator can continue to feed material to the dewatering section. During such periods the thickened underflow should be recirculated into the thickener feedwell. At no time should the underflow cease to be pumped, as choking of the discharge cone rapidly occurs.

7.2.4 Thickener Operation and Control

The two primary functions of the thickener are the production of a clarified overflow and a thickened underflow of the required concentration. For a given throughput, the clarifying capacity is determined by the thickener diameter, since the surface area must be large enough so that the upward velocity of liquid is at all times lower than the settling velocity of the slowest settling particle that is to be recovered. The degree of thickening produced is controlled by the residence time of the particles and hence by the thickener depth.

The solids concentration in a thickener varies from that of the clear overflow to that of the thickened underflow being discharged (Fig. 7.9). When materials settle with a definite interface between the suspension and the clear liquid, as is the case with most flocculated mineral pulps, the solids-handling capacity determines the surface area. Solids-handling capacity is defined as the capacity of a material of given dilution to reach a condition such that the mass rate of solids leaving a region is equal to or greater than the mass rate of solids entering the region. The attainment of this condition with a specific dilution depends on the mass subsidence rate being equal to or greater than the corresponding rise rate of displaced liquid. A properly sized thickener containing material of many different dilutions, ranging from the feed to the underflow solids contents, has adequate area such that the rise rate of displaced liquid at any region never exceeds the subsidence rate.

Operation of the thickener to provide clarified overflow depends upon the existence of a clear-liquid zone at the top. If the zone is too shallow, some of the smaller particles may escape in the overflow. The volumetric rate of flow upwards is equal to the difference between the rate of feed of liquid and the rate of removal in the underflow. Hence the required concentration of solids in the underflow, as well as the throughput, determines the conditions in the clarification zone. Although not normally a problem with clarifiers, a stable froth bed can sometimes form on the surface of thickeners which may hinder operation (overflow is from under the froth bed in these cases). Islands of solids may also form.

Control of thickeners requires certain measurements. Concha (2014) describes novel instrumentation for operations to provide estimates of settling velocity and "solids stress" on the compacted solids. For on-line measurements, the following are considered among the most important.

(1) Bed Level. The ability to monitor the "bed level" in a thickener is crucial in enhancing efficiency (Ferrar, 2014). Incorrect measurements can lead to problems such as excess water reporting to the underflow, sludge spillage into the overflow, or incorrect feedback in control of flocculation. Various techniques are employed depending on the application, including: calculated bed level based on density and hydrostatic pressure; ultrasound transducers to sense reflections from the solid bed; buoyancy-based electromechanical system; and conductivity-based probes.

(2) Feed Mass Flow Rate. This is important to control clarity of the overflow water. Throughput can be optimized by combining mass flow measurement with ratio control of the flocculant dosage.

(3) Flocculant Dosage Rate. As flocculants are expensive, keeping dosage to a minimum consistent with target performance is a priority and is one key to minimizing operating costs (Ferrar, 2014).

7.2.5 Thickener Sizing

Sizing methods include by experience, cylinder (or jar) settling tests, and pilot plant testwork (McIntosh, 2009). Experience (or "rule of thumb") is used if there is no sample, or to prepare a rough first draft for sizing equipment for budget purposes at the beginning of the project. Pilot plant testwork is the most reliable. A number of pilot plant units are available in various sizes and configurations. These units are small scale versions of commercial thickeners, with a feedwell, flocculant addition facility, underflow pumps, and rake mechanisms to duplicate the full-scale thickening process.

Batch settling tests, described above (Section 7.2.1), are most commonly used. There are two well-known methods to design thickeners based on the settling curve, the Coe and Clevenger (1916) method and the Talmage and Fitch (1955) method. The first determines the thickener area as follows.

If X is the liquid-to-solids ratio by weight at any region within the thickener, U the liquid-to-solids ratio of the thickener discharge, and W (t/h) the dry solids feed rate to the thickener, then, assuming no solids leave with the overflow, $(X - U)W$ (t/h) mass of liquid moves upwards with velocity V (m/h):

$$V = \frac{(X - U)W}{AS} \tag{7.1}$$

where A is the thickener area (m²) and S the density of the liquid (t/m³). Because this upward velocity must not exceed the settling rate of the solids in this region, in the limit:

$$R = \frac{(X - U)W}{AS} \tag{7.2}$$

where R is the settling rate (m/h).

The required thickener area (m²) is, therefore:

$$A = \frac{(X - U)W}{AS} \tag{7.3}$$

From a set of R and X values, the area required for various dilutions may be found by recording the initial settling rate as a function of dilution ranging from that of the feed to the discharge. The dilution corresponding to the maximum value of A represents the minimum solids-handling capacity and is the critical dilution. A scale up safety factor of between 1.2 and 1.5 is applied to A (Dahlstrom, 2003).

Originally proposed for suspensions without flocculant, in the Coe-Clevenger method the error in determining the settling rate increases as the flocculant dosage is increased. Parsapour et al. (2014) have proposed a modified procedure to account for the addition of flocculant.

The Coe and Clevenger method requires multiple batch tests at different pulp densities before an acceptable unit area can be selected. The Kynch model (1952) offers a way of obtaining the required area from a single batch-settling curve and is the basis of several thickening theories, which have been comprehensively reviewed by Pearse (1977) and Concha (2014).

The Talmage and Fitch method (1955) applies Kynch's model to the problem of thickener design. They showed that by constructing a tangent to the curve at any point on the settling curve (Fig. 7.13), then:

$$CH = C_0 H_0 \qquad (7.4)$$

where H(cm) is the interface height corresponding to a uniform slurry of concentration C (kg/L) at the point where the tangent was taken, C_0(kg/L) is the original feed solids concentration, and H_0 (cm) the original interface height. Therefore, for any selected point on the settling curve, the local concentration can be obtained from Eq. (7.4),

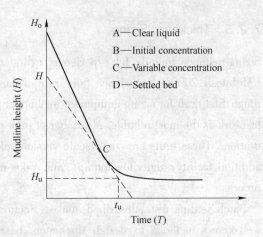

Fig. 7.13 Batch settling curve
(Simplified version of Fig. 7.2)

and the settling rate from the gradient of the tangent at that point. Thus a set of data of concentration against settling rate can be obtained from the single batch-settling curve.

To understand the approach we start by rewriting Eq. (7.3) in terms of concentration C. For a pulp of solids concentration C (kg/L), the volume occupied by the solids in 1 liter of pulp is C/d, where d (kg/L) is the density of dry solids.

Therefore the weight of water in 1 liter of pulp is:

$$1 - \frac{C}{d} = \frac{d-C}{d}$$

and the water-solids ratio by weight become:

$$C_u = \frac{d-C}{dC}$$

For pulps of concentrations C(kg/L) of solids, and C_u(kg/L) of solids, the difference in water-solids ratio is:

$$\text{water-solids ratio} = \frac{d-C}{dC} - \frac{d-C_u}{dC_u} = \frac{1}{C} - \frac{1}{C_u}$$

Therefore, the values of concentration obtained, C, and the settling rates, R, can be substituted in the Coe and Clevenger Eq. (7.3) to give:

$$A = \left(\frac{1}{C} - \frac{1}{C_u} \right) \frac{W}{R} \qquad (7.5)$$

where C_u is the underflow solids concentration. (Note, to preserve A in m^2 C is now in t/m^3 and S is eliminated.)

A simplified version of the Talmage and Fitch method is offered by determining the point on the settling curve where the solids go into compression. This point corresponds to the critical sedimentation point in Fig. 7.2, and controls the area of thickener required. In Fig. 7.13, C is the critical sedimentation point (or compression point) and a tangent is drawn to the curve at this point, intersecting the ordinate at H. A line is drawn parallel to the abscissa corresponding to the target underflow solids concentration C_u which intersects the ordinate at H_u. The tangent from C intersects this line at a time corresponding to t_u.

The required thickener area from Eq. (7.5) is then:

$$A = \frac{W(1/C - 1/C_u)}{(H - H_u)/t_u}$$

where $R = (H/H_u)/t_u$ is the gradient of the tangent at point C, that is, the settling rate of the particles at the compression point concentration. Since $CH = C_0H_0$, then:

$$A = \frac{W[(H/C_0H_0) - (H_u/C_0H_0)]}{(H - H_u)/t_u}$$

That is:

$$A = W\frac{t_u}{C_0H_0} \tag{7.6}$$

In most cases, the compression point concentration will be less than that of the underflow concentration. In cases where this is not so, the tangent construction is not necessary, and tu is the point where the underflow line crosses the settling curve. The point of compression on the curve can be clear, but when this is not so, a variety of methods have been suggested for its determination (Fitch, 1977; Pearse, 1980; Laros et al., 2002).

The Coe and Clevenger and modified Talmage and Fitch methods are the most widely used in the metallurgical industry to predict thickener area requirements. Both methods have limitations (Waters and Galvin, 1991; Parsapour et al., 2014): the Talmage and Fitch technique relying critically on identifying a compression point, and both methods must be used in conjunction with empirical safety factors. However, the results of these methods are similar when the settling tests are carried out either on a single sample with the solids concentration of the thickener feed using the Talmage and Fitch method or on diluted samples using the Coe and Clevenger method (Parsapout et al., 2014).

Software has been developed for prediction of thickener area based on a phenomenological model of particle settling. The development of thickener models is reviewed by Concha and Burger (2003) and Concha (2014).

The mechanism of solids consolidation has been far less well expressed in mathematical terms than the corresponding clarifying mechanisms. The height of the thickener is, therefore, usually determined by experience.

7.2.6 Other Gravity Sedimentation Devices

7.2.6.1 Tray Thickener

The diameter of a conventional thickener is usually large and therefore a large ground area is required. Tray thickeners (Fig. 7.14) are sometimes installed to save space. In essence, a tray thickener is a series of unit thickeners mounted vertically above one another. They operate as separate units, but a common central shaft is utilized to drive the sets of rakes.

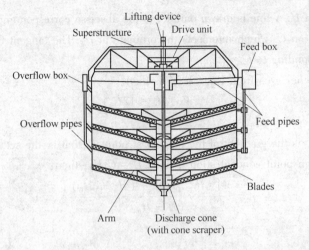

Fig. 7.14 Tray thickener

7.2.6.2 Lamella Thickener

Also known as an inclined plate settler, it has two main parts, an upper tank containing lamella plates inclined at 55° and a lower conical or cylindrical tank with a rake mechanism (Fig. 7.15). The inclined plate gives a short distance for particle sedimentation, and with low friction, sliding down the plate increases the speed of separation (Anon., 2011). Clarification (clear supernatant) is achieved when the upstream liquid velocity is sufficiently low to allow solids to settle to the plate. Thickening is achieved through a combination of sedimentation onto the plate ("primary thickening") and conventional sedimentation in the lower tank.

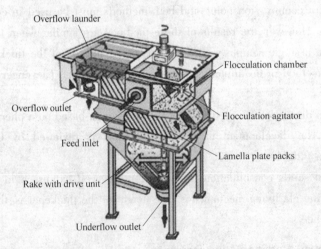

Fig. 7.15 Lamella thickener/clarifier

7.3 CENTRIFUGAL SEDIMENTATION

Centrifugal separation can be regarded as an extension of gravity separation, as the settling rates of particles are increased under the influence of centrifugal force. It can, however, be used to separate emulsions which are normally stable in a gravity field.

Centrifugal separation can be performed either by hydrocyclones or centrifuges. The simplicity and cheapness of the hydrocyclone (Chapter 3) make it very attractive, although it suffers from restrictions with respect to the solids concentration that can be achieved and the relative proportions of overflow and underflow into which the feed may be split. Generally, the efficiency of even a small-diameter hydrocyclone falls off rapidly at very fine particle sizes and particles smaller than about 10μm in diameter will invariably appear in the overflow, unless they have high density. Flocculation of such particles is limited, since the high shear forces due to the cyclonic action break up the agglomerates. The hydrocyclone is therefore inherently better suited to classification rather than thickening. Its dry counterpart, the (air) cyclone, is widely used for "dust" removal in a variety of industrial applications.

By comparison, centrifuges are much more costly and complex, but have a much greater clarifying power and are generally more flexible than hydrocyclones. Various types of centrifuge are used industrially (Bragg, 1983; Bershad et al. , 1990; Leung, 2002), the solid bowl centrifuge (or decanter) having widest use in the minerals industry due to its versatility and ability to discharge the solids continuously.

The basic principles of a typical centrifuge are shown in Fig. 7.16. It consists of a horizontal revolving shell or bowl, cylindroconical in form, inside which a screw conveyor of similar section rotates in the same direction at a slightly higher or lower speed. The feed pulp is admitted to the bowl through the center tube of the revolvingscrew conveyor. On leaving the feed pipe, the slurry is immediately subjected to a high centrifugal force, causing the solids to settle on the inner surface of the bowl at a rate which depends on the rotational speed employed, this normally being between 1600 and 8500r/min. The separated solids are conveyed by the scroll out of the liquid and discharged through outlets at the smaller end of the bowl. The solids are continuously dewatered by centrifugal force as they proceed from the liquid zone to the discharge. Excess entrained liquor drains away to the pond circumferentially through the particle bed. When the liquid reaches a predetermined level, it overflows through the discharge ports at the larger end of the bowl.

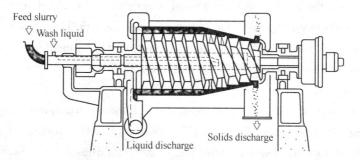

Fig. 7.16 Continuous solid bowl centrifuge

The actual size and geometry of these centrifuges vary according to the throughput required and the application.

The length of the cylindrical section largely determines the clarifying power and is thus made a maximum where overflow clarity is of prime importance. The length of the conical section, or "beach", decides the residual moisture content of the solids, so that a long shallow cone is used where maximum dryness is required.

Centrifuges are manufactured with bowl diameters ranging from 15 to 150cm, the length generally being about twice the diameter. Throughputs vary from about 0.5 to 50m^3/h of liquid and from about 0.25 to 100t/h of solids, depending on the feed concentration, which may vary widely from 0.5% to 70% solids, and on the particle size, which may range from about 12mm to as fine as 2μm, or even less when flocculation is used. The application of flocculation is limited by the tendency of the scroll action to damage the flocs and thus redisperse the fine particles. The moisture content in the product varies widely, typically being in the range 5%~20%.

7.4 FILTRATION

Filtration is the process of separating solids from liquid by means of a porous medium (the filter) which retains the solid but allows the liquid to pass. The most common filter type in mineral processing is cake filtration, where the liquid passes through the filter, called the filtrate, and the solids build-up on the filter is referred to as filter cake. The volume of filtrate collected per unit time is the rate of filtration. There are typically five steps in the process: cake formation, moisture reduction, cake washing (if required), cake discharge, and medium washing.

Filtration in mineral processing applications normally follows thickening. The thickened pulp may be fed to storage agitators from where it is drawn off at uniform rate to the filters. Flocculants are sometimes added to the agitators in order to aid filtration. Slimes have an adverse effect on filtration, as they tend to "blind" the filter medium; flocculation reduces this and increases the voidage between particles, making filtrate flow easier. The lower molecular weight flocculants tend to be used in filtration, as the flocs formed by high molecular weight products are relatively large, and entrain water within the structure, increasing the moisture content of the cake. With the lower molecular weight flocculants, the filter cake tends toward a uniform porous structure which allows rapid dewatering, while still preventing migration of fine particles through the cake (Moss, 1978). Other surfactant filter aids are used to reduce the liquid surface tension, or more likely, to modify particle surface properties to assist flow through the medium (Singh et al., 1998; Wang et al., 2010; Asmatulu and Yoon, 2012).

There is a large body of literature on types of filters, filtration principles, equipment selection, testing and sizing (Wakeman and Tarleton, 1999; Svarovsky, 2000; Rushton et al., 2000; Smith and Townsend, 2002; Cox and Traczyk, 2002; Welch, 2002; Dahlstrom, 2003; Stickland, 2008; Concha, 2014).

7.4.1 Brief Theory

Concha (2014) provides comprehensive theoretical treatment; the brief summary here based on

Smith and Townsend (2002), Cox and Traczyk (2002), and Dahlstrom (2003) is just to introduce the main operational variables. Starting with the classical theory of Darcy and Poiseuille, the basic filtration equation can be written as:

$$V = \frac{1}{A}\frac{dV}{dt} = \frac{\Delta P}{\mu\left(\alpha w \dfrac{v}{A}\right)} \tag{7.7}$$

where v is filtrate flow rate (m/s), A area of filter (m^2), V filtrate volume (m^3), t time (s), ΔP pressure drop across the cake and medium (N/m^2), μ liquid viscosity (N·s/m^2), α specific cake resistance (m/kg), and w feed slurry concentration in terms of dry solids mass per unit of filtrate volume (kg/m^3). The equation may be modified by adding a resistance term for the filter medium. Eq. (7.7) shows that the filtration rate varies directly with the pressure drop across the filter and the area of the filter, and inversely with liquid viscosity, cake resistance (reciprocal of cake permeability), and slurry solids contents. These factors determine the necessary test variables in sizing filters.

The cake comprises a bundle of capillaries and water can be removed only when the applied pressure is greater than the capillary pressure P_C given by:

$$P_C = \frac{4\gamma_l \cos\theta}{d_C} \tag{7.8}$$

where γ_l is the liquid surface tension, θ the contact angle, and d_C is the diameter of the capillary. Eq. (7.8) indicates why surfactant filter aids can assist, by decreasing surface tension and/or increasing contact angle (Asmatulu and Yoon, 2012). The equation also indicates that filtration becomes more difficult as particle size decreases, as the capillary (pore) diameter tends to decrease.

7.4.2 The Filter Medium

The choice of the filter medium is often the most important consideration in assuring efficient operation of a filter. Its function is generally to act as a support for the filter cake, while the initial layers of cake provide the true filter. The filter medium should be selected primarily for its ability to retain solids without blinding. It should be mechanically strong, corrosion resistant, and offer as little resistance to flow of filtrate as possible. Relatively coarse materials are normally used and clear filtrate is not obtained until the initial layers of cake are formed, the initial cloudy filtrate being recycled.

Filter media are manufactured from cotton, wool, linen, jute, silk, glass fiber, porous carbon, metals, rayon, nylon and other synthetics, ceramic, and miscellaneous materials such as porous rubber. Cotton fabrics are among the most common type of medium, primarily because of their low initial cost and availability in a wide variety of weaves.

7.4.3 Filtration Tests

It is not normally possible to forecast what may be accomplished in the filtration of an untested product, therefore preliminary tests have to be made on representative samples of pulp before the large-scale plant is designed. Bench scale testing of samples for specification of filtration equipment is described by Smith and Townsend (2002) and Tarleton and Wakeman (2006). Tests are also commonly carried out on pulps from existing plants, to assess the effect of changing operating conditions, filter aids, etc.

It is necessary to identify what are the objectives, for example, target moisture, and what are the candidate filter types, for example, pressure or vacuum? The slurry physical and chemical conditions that need to be replicated in the sample should include: solids concentration, particle density and size distribution, slurry pH, and chemical additives (e. g., flotation reagents and flocculants).

7.4.3.1 Vacuum Filtration Test

A simple vacuum filter leaf test setup is shown in Fig. 7.17. The filter leaf, consisting of a section of the industrial filter medium, is connected to a filtrate receiver equipped with a vacuum gauge. A known weight of slurry is introduced sufficient to approximate the target cake thickness. The receiver is connected to a vacuum pump. If the industrial filter is to be a continuous vacuum filter, this operation must be simulated in the test. The cycle is divided

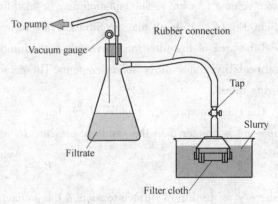

Fig. 7.17 Laboratory test filter

into three sections: cake formation (or "pick-up"), drying, and discharge. Sometimes pick-up is followed by a period of washing and the cake may also be subjected to compression during drying. While under vacuum, the test leaf is submerged for the pick-up period in the agitated pulp to be tested. The leaf is then removed and held with the drainpipe down for the allotted drying time.

At the end of drying, the filter cake is removed and the net weight and cake thickness are recorded. The sample is dried and weighed to determine moisture content. The daily filter capacity can then be determined by the dry weight of cake per unit area of test leaf multiplied by the daily number of cycles and the filter area. A range of conditions should be tested to cover the range of anticipated variables.

7.4.3.2 Pressure Filtration Test

Bench-scale pressure filtration testwork can be performed using a bomb device. A typical apparatus is a 250mm length of 50mm (outer-wall diameter) pipe capped with flanges. The lower flange supports the test filter cloth on a drainage grid above the filtrate collection port. The upper flange houses the air pressure connection, pressure gauge, and feed port. The test procedure is similar to that described above, with applied pressure being substituted for vacuum. A new pressure test procedure, step pressure filtration, suited to flocculated feeds, is described by Usher et al. (2001) and De Kretser et al. (2011).

7.4.4 Types of Filter

For particles coarse enough that capillary pressures are negligible (Eq. (7.8)), gravimetric dewatering can be employed, for example, dewatering screens in dense media recovery (Chapter 5). This is not usually the case and cake filters are the type most frequently used in mineral processing, where the recovery of large amounts of fine solids (typically < 100 μm) from fairly concentrated slurries (50% ~ 60% solids) is the main requirement. Cake filters may be pressure or vacuum types, and operation may be batch or continuous (Cox and Traczyk, 2002). In pressure filters positive pressure is applied at the feed end and in vacuum filters there is a vacuum at the far side of the filter, the feed side being at atmospheric pressure. Dewatering is a combination of cake compression and air blow through.

7.4.4.1 Pressure Filters

Because of the increasing fineness of mineral concentrates (those of Cu, Pb, and Zn are commonly 80% < 30 μm), coupled with shipping schedules calling for moisture contents 8wt% ~ 10wt% on these fine concentrates, filtration under pressure has certain advantages over vacuum. (Given that many operations are at high altitude is an additional drawback for vacuum units.) Higher flow rates and better washing and drying result from the higher pressures that can be used. Pressure filters have become sufficiently large and reliable to handle the output of most concentrators and can produce low enough cake moisture to eliminate driers (Townsend, 2003). Thus, the trend is to pressure filtration (Cox and Traczyk, 2002).

The common pressure filters in mineral processing applications come in two basic forms, horizontal and vertical, defined either by the orientation of the filter plates (Concha, 2014), or the convention here, by the direction the pressure is applied: actuation either horizontally or vertically (Cox and Traczyk, 2002). They both represent more automated versions of plate-and-frame filters (Taggart, 1945; McCaslin et al., 2014).

7.4.4.2 Horizontal Pressure Filters

A typical horizontal filter is shown in Fig. 7.18. In this arrangement, filter plates, usually made from

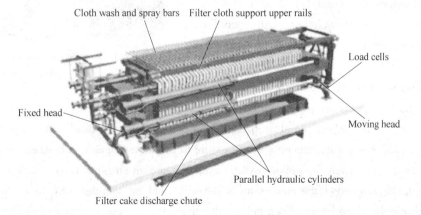

Fig. 7.18 Horizontal pressure filter (Courtesy Metso)

lightweight polymer, are suspended vertically from a steel frame (hence the alternative name vertical plate pressure filter). Between the plates is hung filter cloth. The plates are held together and the press is opened and closed by a hydraulic piston. The slurry is pumped into the press to fill each chamber. Dewatering starts immediately the slurry enters, and the filtrate is removed. When the chamber is full of material, membranes on one side of the chamber are pressurized to hold the cake in place and squeeze out some water. At the end of this "feeding cycle", the main dewatering cycle is activated by forcing pressurized air through the cake, the "air dewatering cycle" with pressures up to 8 bar. A "washing cycle" could be incorporated, the cake washed by replacing air by liquid with air pressure then re-applied to achieve final moisture.

At completion, the press is opened, the plates separate and the cake falls by gravity onto a conveyor. To finish, the filter cloth is cleaned by a combination of vibration and water sprays, the washings recycled to the feed tank, the chambers closed and the cycle repeated. While the operation is batch, with cycle times of ca. 10min and a feed tank with sufficient storage, operation appears continuous.

7.4.4.3 Vertical Pressure Filters

These differ from the horizontal units by stacking the chambers on top of each other and rather than individual filter cloths between each chamber, the cloth in the vertical unit is continuous (Fig. 7.19). The filtration cycle is similar to that for horizontal units. At completion, the press is opened and the filter cloth advanced to discharge the cake, followed by cloth washing. Both horizontal and vertical units are automated to control the various cycle times and may include sensors to monitor, for example, cloth condition (Townsend, 2003).

7.4.4.4 Tube Press

In some situations, dewatering of ultrafine ($<10\mu m$) material requires special equipment. Large capillary resistance forces demand greater air pressures (Eq. (7.8)) than in the units described above. By performing filtration in a tube, pressures up to 100bar can be

Fig. 7.19 Vertical pressure filter
(Courtesy FLSmith)

exploited. The tube press has been applied in a variety of difficult dewatering applications. It consists of a casing with a membrane at each end and a porous tube (or "candle") covered with cloth suspended inside. Feed is introduced under pressure to fill the casing and cake starts to form on the candle. Dewatering pressure is applied by the membrane and an air blow. Cake washing can also be incorporated. On completion, the membrane is retracted, and the candles lowered to discharge the cake, which can be aided by air blown behind the cloth. The candle is reinserted in the casing and the cycle repeats. A related device is the "candle filter", which consists of a series of porous tubes (candles) inside a pressurized chamber (Concha, 2014) (Fig. 7.20).

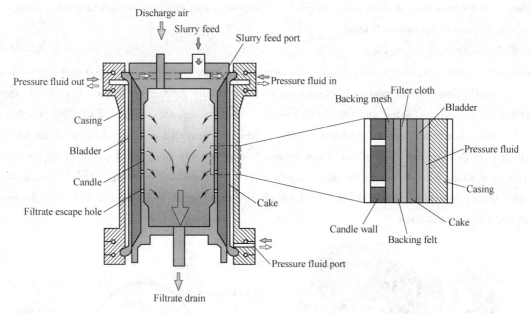

Fig. 7.20 Operation of tube press

7.4.4.5 Vacuum Filters

There are many different types of vacuum filter, but they all incorporate filter media suitably supported on a drainage system, beneath which the pressure is reduced by connection to a vacuum system.

7.4.4.6 Batch Vacuum Filters

There are two main types, the vertical leaf filter and the horizontal leaf or tray filter, which are similar except for the orientation of the leaf. The leaf consists of a metal framework or a grooved plate over which the filter cloth is fixed (Fig. 7.21). Numerous holes are drilled in the pipe framework, so that when a vacuum is applied, a filter cake builds up on both sides of the leaf. A number of leaves are generally connected. For example, in the vertical leaf filter, the array is first immersed in slurry held in a slurry feed tank, removed, and then placed in a cakereceiving tank where the cake is removed by replacing the vacuum by air pressure. Although simple to operate, these filters require considerable floor space. They are now used only for clarification, that is, the removal of small amounts of suspended solids from liquors.

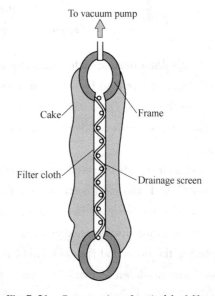

Fig. 7.21 Cross section of typical leaf filter

7.4.4.7 Continuous Vacuum Filters

These are the most widely used filters in mineral processing applications and fall into three classes: drums, discs, and horizontal filters.

7.4.4.8 Rotary-Drum Filter

This is the most common, finding application both where cake washing is required and where it is unnecessary. The drum is mounted horizontally and is partially submerged in the filter trough, into which the feed slurry is fed and maintained in suspensionby agitators (Fig. 7.22). The periphery of the drum is divided into compartments, each of which is provided with a number of drain lines, which pass through the inside of the drum, terminating at one end as a ring of ports, which are covered by a rotary valve to which vacuum is applied. The filter medium is wrapped tightly around the drum surface which is rotated at low speed, usually in the range 0.1~0.3r/min, but up to 3r/min for fast filtering materials.

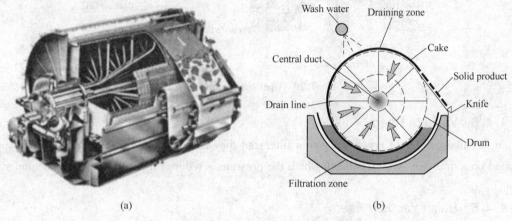

Fig. 7.22 Rotary-drum filter with belt discharge
(a) cut-away diagram, and (b) schematic cross-section

As the drum rotates, each compartment goes through the same cycle of operations, the duration of each being determined by the drum speed, the depth of submergence of the drum, and the arrangement of the valve. The normal cycle of operations consists of filtration, drying, and discharge, but it is possible to introduce other operations into the basic cycle, such as cake washing and cloth cleaning.

Various methods are used for discharging the solids from the drum, depending on the material being filtered. The most common form makes use of a reversed blast of air, which lifts the cake so that it can be removed by a knife, without the latter actually contacting the medium. Another method is string discharge, where a number of endless strings around the drum lift the filter cloth as it leaves the drum and the cake falls off. It is rarely used today. An advance on this method is belt discharge, as shown in Fig. 7.22, where the filter medium itself leaves the filter and passes over an external roller, before returning to the drum. This has a number of advantages in that very much thinner cakes can be handled, with consequently increased filtration and draining rates and hence better washing and dryer products. At the same time, the cloth can be washed on both sides by means of sprays before it returns to the drum, thus minimizing the extent of blinding. Cake washing is usually carried out by means of sprays or weirs, which cover a fairly limited area at the top of the drum.

The capacity of the vacuum pump will be determined mainly by the amount of air sucked through the cake during the washing and drying periods when, in most cases, there will be a simultaneous flow of both liquid and air. A typical layout is shown in Fig. 7.23, from which it is seen that the air and liquid are removed separately. The barometric leg should be at least 10m high to prevent liquid being sucked into the vacuum pump.

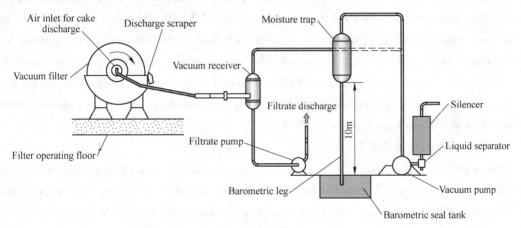

Fig. 7.23 Typical rotary-drum filter system

Variations on standard drum filters to enable them to handle coarse, free-draining, quick-settling materials include top feed units where the material is distributed at between 90° and 180° from the feed point.

7.4.4.9 Disc Filter

The principle of operation of disc filters (Fig. 7.24), is similar to that of rotary drum filters. The disc filter consists of sectors of cloth covered steel mounted on a central shaft that also connects a

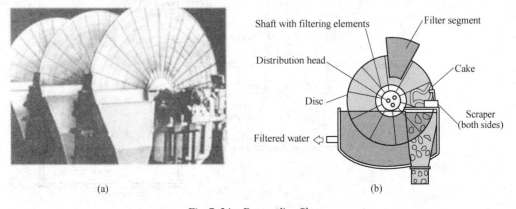

Fig. 7.24 Rotary-disc filter:
(a) general view, and (b) schematic cross-section

certain number of the sectors to vacuum. The solids cake is formed on both sides of the disc; the disc rotates and lifts the cake above the level of the slurry in the trough, whereupon the cake is suction-dried and is then removed by a pulsating air blow with the assistance of a scraper. Several discs are mounted along the shaft separated by about 30cm and consequently a large filtration area can be accommodated in a small floor space. Cost per unit area is thus lower than for drum filters, but cake washing is virtually impossible with the disc filter.

7.4.4.10 Ceramic Disc Filter

A special type of disc filter uses micro-pore ceramic sectors rather than ones of steel covered with cloth. When submerged in the slurry pool, capillary action assists drawing liquid through the pores of the filter. This can be understood by reference to Eq. (7.8); for the filter (as opposed to the cake) we want the capillary pressure to be high, achieved by having small pore diameter and strongly hydrophilic material, that is, ceramic. The action reduces the size of vacuum pump required, resulting in reduced energy consumption. Referred to as capillary filtration, it can produce moisture contents that approach pressure filtration and has found application on mineral concentrates (Cox and Traczyk, 2002; Concha, 2014).

7.4.4.11 Horizontal Belt Filter

If ultimate cake moisture is not critical, but rather water recovery or production of solids that can be handled or stored is critical, then the belt filter may be suitable (Fig. 7.25). Examples include tailings dewatering at operations with limited space or environmental restrictions on disposal in tailings ponds, and recovery of leach liquors in hydrometallurgical operations. It comprises an endless

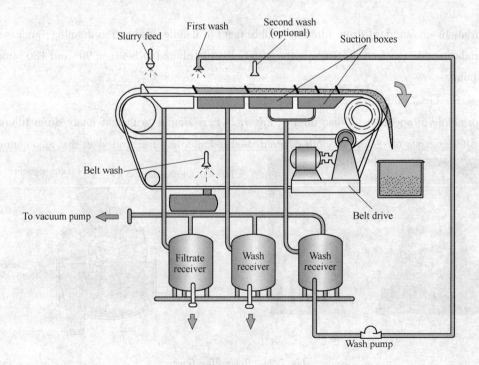

Fig. 7.25 Horizontal belt filter

perforated rubber drainage deck supporting the filter cloth. Vacuum is applied by a series of suction boxes underneath the belt. Varying the number of the suction boxes gives control over the length of the filtration, drying, and washing (if included) stages. The cake is discharged as the belt reverses over a small diameter roller.

7.4.4.12 Pan Filter

This consists of a series of horizontal trays supporting the filter cloth rotating around a central vertical axis and connected to a common suction valve. The trays are trapezoidal in shape to accommodate the rotation and slightly tilted toward the center. Cake builds, can be washed, and is discharged by tipping the tray. Compared to a disc filter, which has two sides forming cake, the pan filter, with only one side, is lower capacity.

7.4.4.13 Hyperbaric Filters

By placing a conventional disc or drum vacuum filter inside a pressurized vessel, the available pressure drop can be increased, up to four bars or more (Bott et al., 2003; Concha, 2014). Cake moisture levels can be reduced from typically 15% in vacuum filters to ca. 8%. As with all pressure filters, cake discharge is a problem.

7.5 DRYING

The drying of concentrates prior to shipping, if done, is the last operation performed in the mineral processing plant. It reduces the cost of transport and is usually aimed at lowering the moisture content to about 5wt%. Dust losses are often a problem if the moisture content is lower.

Prokesch (2002), Mujumdar (2006), and Kudra and Mujumdar (2009) review the types of drying equipment available and describe dryer selection based on the required duty. The dryer types include hearth, grate, shaft, fluidized bed, and flash. The one often used in concentrators is the rotary dryer.

Rotary Thermal Dryer: This unit consists of a relatively long cylindrical shell mounted on rollers and driven at a speed of up to 25r/min. The shell is at a slight slope, so that material moves from the feed to discharge end under gravity. Hot gases, or air, are fed in either at the feed end to give parallel flow or at the discharge to give counter-current flow.

The method of heating may be either direct, in which case the hot gases pass through the material in the dryer, or indirect, where the material is in an inner shell, heated externally by hot gases. The direct-fired is the one most commonly used in the minerals industry, the indirect-fired type being used when the material must not contact the hot combustion gases. Parallel flow dryers (Fig. 7.26) are used in the majority of operations because they are more fuel efficient and have greater capacity than counterflow types (Kram, 1980). Since heat is applied at the feed end, build-up of wet feed is avoided, and in general these units are designed to dry material to not less than 1% moisture. Since counter-flow dryers apply heat at the discharge end, a completely dry product can be achieved, but its use with heat-sensitive materials is limited because the dried material comes into direct contact with the heating medium at its highest temperature.

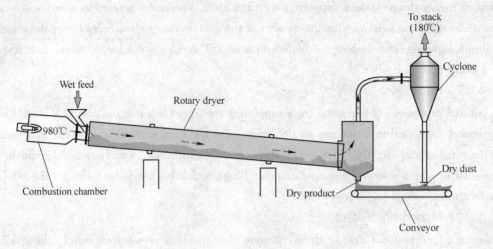

Fig. 7.26 Direct fired, parallel flow rotary dryer
(Adapted from Kram (1980))

The product from the dryers is often stockpiled, before being loaded on to trucks or rail-cars as required for shipment. To control dust, containers may be closed, or the surface of the contents sprayed with various dust suppressing solutions (Kolthammer, 1978).

REFERENCES

Anon, 2011. Basics in Mineral Processing. Eighth ed. Metso Corporation. Arbuthnot, I., et al., 2005. Designing for paste thickening. Proc. 37th
  Annual Meeting of Canadian Mineral Processors Conf. CIM, Ottawa, ON, Canada, 597-628.

  Asmatulu, R., Yoon, R. H., 2012. Effects of surface forces on dewatering of fine particles. In: Young, C. A., Luttrell, G. H. (Eds.), Separation Technologies for Mineral, Coal, and Earth Resources. SME, Englewood, CO, USA, 95-102.

  Bershad, B. C., et al., 1990. Making centrifugation work for you. Chem. Eng. 97 (7-12), 84-89.

  Bott, R., et al., 2003. Recent developments and results in continuous pressure and steam-pressure filtration. Aufbereitungs Techni. 44, 518.

  Bragg, R., 1983. Filters and centrifuges (part 1- part 4). Mining Mag. Aug., 90-111.

  Bulatovic, S. M., 2007. Handbook of Flotation Reagents, Theory and Practice. Flotation of Sulfide Ores, vol. 1. Elsevier, Amsterdam, Netherlands.

  Buscall, R., White, R. L., 1987. The consolidation of concentrated suspensions. Part 1. The theory of sedimentation. J. Chem. Soc., Faraday Trans. 1: Phys. Chem. Condensed Phases. 83 (6), 873-891.

  Coe, H. S., Clevenger, G. H., 1916. Methods for determining the capacities of slime-settling tanks. Trans. AIMME. 55, 356-384.

  Concha, F., Burger, R., 2003. Thickening in the 20th century: a historical perspective. Miner. Metall. Process. 20 (2), 57-67.

  Concha, F. A., 2014. Solid-Liquid Separation in the Mining Industry, vol. 105. Springer, Fluid Mechanics and Its Applications (Series).

  Cox, C., Traczyk, F., 2002. Design features and types of filtration equipment. In: Mular, A. L., et al., (Eds.),

Mineral Processing Plant Design, Practice and Control, vol. 2. SME, Littleton, CO, USA, 1342-1357.

Dahlstrom, D. A. , 2003. Liquid-solid separation. In: Fuerstenau, M. C. , Han, K. N. (Eds.) , Principles of Mineral Processing. SME, Littleton, CO, USA, 307-362.

De Kretser, R. G, et al. , 2011. Comprehensive characterisation of material properties for dewatering: how much is enough? FILTECH 2011 Congress Proc. , vol. 1. Wiesbaden, Germany, 383-390.

Farrow, J. B. , Swift, J. D. , 1996. A new procedure for assessing the performance of flocculants. Int. J. Miner. Process. 46 (3-4) , 263-275.

Ferrar, G. , 2014. Optimise your thickener efficiency for maximum profitability. <http://www.pacetoday.com.au/news/optimise-your-thickener-efficiency-for-maximum-pro>. (Viewed December 2014).

Fitch, E. B. , 1977. Gravity separation equipment-clarification and thickening. In: Purchas, D. B. , Wakeman, R. J. (Eds.), Solid-Liquid Separation Equipment Scale-up. Uplands Press, Croydon, UK. Hogg, R. , 2000. Flocculation and dewatering. Int. J. Miner. Process. 58 (1-4) , 223-236.

Hsia, E. S. , Reinmiller, F. W. , 1977. How to design and construct earth bottom thickeners. Mining Eng. Aug. , 36-39.

Kolthammer, K. W. , 1978. Concentrate drying, handling and storage. In: Mular, A. L. , Bhappu, R. B. (Eds.), Mineral Processing Plant Design. AIMME, New York, NY, USA, 601-617.

Kram, D. J. , 1980. Drying, calcining, and agglomeration. Eng. Min. J. 181 (6) , 134-151.

Kudra, T. , Mujumdar, A. S. , 2009. Advanced drying technologies. Seconded. CRC Press, Taylor & Francis Group, Boca Raton, FL, USA.

Kurt, N. , 2006. A Study of Channelling Behaviour in Batch Sedimentation. PhD thesis. School of Civil and Chemical Engineering, Royal Melbourne Institute of Technology University, Melbourne, Australia.

Kynch, G. J. , 1952. A theory of sedimentation. Trans. Faraday Soc. 48, 166-176.

Lake, P. , Summerhays, R. P. , 2012. Thickener feed system design. Proc. 13th International Mineral Processing Symp. Bodrum, Turkey, 197-204.

Laros, T. , et al. , 2002. Testing, sizing, and specifying sedimentation equipment. In: Mular, A. L. , et al. , (Eds.), Mineral Processing Plant Design, Practice and Control, vol. 2. SME, Littleton, CO, USA, 1295-1312.

Leung, W. , 2002. Centrifugal sedimentation and filtering for mineral processing. In: Mular, A. L. , et al. , (Eds.), Mineral Processing Plant Design, Practice and Control, vol. 2. SME, Littleton, CO, USA, 1262-1288.

Loan, C. , et al. , 2009. Operational results from the Vane Feedwell-Cutting-Edge modeling turned into reality. Proc. 10th Mill Operators' Conf. AusIMM, Adelaide, SA, Australia, 261-266.

Luttrell, G. H. , 2014. Innovations in coal processing. In: Anderson, C. G. , et al. , (Eds.), Mineral Processing and Extractive Metallurgy: 100 Years of Innovation. SME, Englewood, CO, USA, 277-296.

McCaslin, M. L. , et al. , 2014. Innovations in liquid/solid separation for metallurgical processing. In: Anderson, C. G. , et al. , (Eds.), Mineral Processing and Extractive Metallurgy: 100 Years of Innovation. SME, Englewood, CO, USA, 333-343.

McIntosh, A. , 2009. Thickener sizing and the importance of testwork. Output Aust. Minerals Metals Process. Solut. 23, 9-12.

Moss, N. , 1978. Theory of flocculation. Mine Quarry. 7 (May) , 57-61.

Mujumdar, A. S. , 2006. Handbook of Industrial Drying. Third ed. CRC press, Taylor & Francis Group, Boca Raton, FL, USA.

Mular, A. L. , et al. , 2002. Mineral Processing Plant Design, Practice, and Control, vol. 2. SME, Littleton, CO, USA (Chapter 9).

Owen, A. T. , et al. , 2002. The impact of polyacrylamide flocculant solution age on flocculation performance.

Int. J. Miner. Process. 67 (1-4), 123-144.

Owen, T. A., et al., 2009. The effect of flocculant solution transport and addition conditions on feedwell performance in gravity thickeners. Int. J. Miner. Process. 93 (2), 115-127.

Parsapour, Gh. A. et al., 2014. Effect of settling test procedure on sizing thickeners. Sep. Purif. Technol. 122, 87-95.

Pearse, M. J., 1977. Gravity Thickening Theories: A Review. Department of Industry, Warren Spring Laboratory.

Pearse, M. J., 1980. Factors affecting the laboratory sizing of thickeners.

Plumtree, A. (Ed.), Fine Particle Processing, vol. 2, 1619-1642

Pearse, M. J., 2003. Historical use and future development of chemicals for solid-liquid separation in the mineral processing industry. Miner. Eng. 16 (2), 103-108.

Prokesch, M. E., 2002. Selection and sizing of concentrate drying, handling and storage equipment. In: Mular, A. L., et al., (Eds.), Mineral Processing Plant Design, Practice and Control, vol. 2. SME, Littleton, CO, USA, 1463-1477.

Rushton, A, et al., 2000. Solid-Liquid Filtration and Separation Technology. VCH, Weinheim, Federal Republic of Germany.

Scales, P. J., et al., 2015. Compressional dewatering of flocculated mineral suspension. Can. J. Chem. Eng. 93 (3), 549-552.

Schoenbrunn, F., Laros, T., 2002. Design features and types of sedimentation equipment. In: Mular, A. L., et al., (Eds.), Mineral Processing Plant Design, Practice and Control, vol. 2. SME, Littleton, CO, USA, 1331-1341.

Seifert, J. A., Bowersox, J. P., 1990. Getting the most out of thickeners and clarifiers. Chem. Eng. 97 (7-12), 80-83.

Singh, B. P., et al., 1998. Use of surfactants to aid dewatering of fine clean coal. Fuel. 77 (12), 1349-1356.

Smith, C. B., Townsend, I. G., 2002. Testing, sizing and specifying of filtration equipment. In: Mular, A. L., et al., (Eds.), Mineral Processing Plant Design, Practice and Control, vol. 2. SME, Littleton, CO, USA, 1313-1330.

Somasundaran, P., Wang, D., 2006. In: Wills, B. A. (Series Ed.), Developments in Mineral Processing. Solutions Chemistry: Minerals and Reagents, vol. 17.

Elsevier, Amsterdam, Boston, USA. Stickland, A. D., et al., 2008. Numerical modelling of flexible-membrane plate-and-frame filtration: formulation, validation and optimization. AIChE J. 54 (2), 464-474.

Svarovsky, L., 2000. Gravity clarification and thickening, Solid-Liquid Separation. Fourth. Butterworth-Heinemann, Oxford, UK, 166-190.

Taggart, A. F., 1945. Handbook of Mineral Dressing, Ore and Industrial Minerals. John Wiley, & Sons. Chapman & Hall, Ltd, London, UK.

Talmage, W. P., Fitch, E. B., 1955. Determining thickener unit areas. Ind. Eng. Chem. 47 (1), 38-41.

Tarleton, E. S., Wakeman, R. J., 2006. Solid/liquid Separation: Equipment, Selection and Process Design. Elsevier, Amsterdam, Netherlands.

Townsend, I., 2003. Automatic pressure filtration in mining and metallurgy. Miner. Eng. 16 (2), 165-173.

Tripathy, T., De, B. R., 2006. Flocculation: a new way to treat the waste water. J. Phys. Sci. 10, 93-127.

Usher, S. P., et al., 2001. Validation of a new filtration technique for dewaterability characterization. AIChE J. 47 (7), 1561-1570.

Usher, S. P., et al., 2009. Theoretical analysis of aggregate densification: impact on thickener performance. Chem. Eng. J. 151 (1-3), 202-208.

Wakeman, R. J., Tarleton, E. S., 1999. Filtration: Equipment Selection, Modelling and Process Simulation. First ed. Elsevier Advanced Technology, Elsevier Science Ltd, Oxford, UK.

Wang, X. Y, et al. , 2010. Polymer aids for settling and filtration of oil sands tailings. Can. J. Chem. Eng. 88 (3), 403-410.

Waters, A. G. , Galvin, K. P. , 1991. Theory and application of thickener design. Filtr. Sep. 28 (2), 110-116.

Welch, D. G, 2002. Characterization of equipment based on filtration principals and theory. In: Mular, A. L. , et al. , (Eds.), Mineral Processing Plant Design, Practice and Control, vol. 2. SME, Littleton, CO, USA, 1289-1294.

Vocabulary

A

abietic 松香
abrasion 磨蚀，耐磨，磨损
abrasion index 磨损指数
abscissa 横坐标
abundance 丰富，多，大量
academic 学校的，文学的
accentuate 强调，着重指出
accommodate 容纳，调节，适应，供给
acqnthite 螺状硫银矿
acidity 酸性
acrylate 丙烯酸盐（酯）
activating 活动的，活化的
activation 活动，活化，致活，激活
activator 活化剂
acute 尖锐的，锐角的
adhesion 黏附，黏合，结合
adiabatic 绝热的，不传热的
adventitious 偶然的，非典型的，外来的，不定的
aeetate 乙酸盐丙酯，乙酸纤维素
affinity 结构相似，关系密切
agent 药剂，因素，力量
agglomcratc 成团的，结块的，聚集，（使）成团
aggregate 盘聚集的，聚集体，结合，共计
agitation 搅动，摇动
aguilarite 辉硒铁矿
aikinite 针硫铋铅矿
air lift 空中补给，空运，气升
air-dried 风干的，空气干燥的
air-tight 密闭的
airlift 空运，空气提升，气动提升机，空气开液气
akaganeite 正方纤铁矿
alabandite 硫锰矿
alaskaite 银辉铅铋矿
albite 钠长石
algodonite 微晶砷铜矿
ambient 周围的，环绕四周的
amine 胺

ammoniacal 氨的，氨性的
ammoniacal liquor 氨水
ammonium 铵（基）
amphibole 角闪石
amply 广大（泛）的，充足地，十分，详细地
amyl 戊（烷）基
anaerobic 厌氧的，无氧的
angle of nip 咬入角
anhydrous 无水的
anionic 阴（负）离子的，阴离子型
anivil 铁砧
ankerre 铁白云母
anomaly 不规则，反常，例外，特殊
anorthite 钙长石
anthracites 无烟煤系
apatite 磷灰石
aperture 孔隙，洞，孔
apparatus 一套器械，装置，仪器
apparent 显然的，明白的
appreciable 可看见的，可测量的
aqueous 水的，似水的
aragonite 霰石
arbitrarily 独裁地，专制地，武断地，任意地，人为地
argon 氩气
arguably 可论证地，可辩论地
arrest point 驻点
arrested crushing 夹压破碎
arsenic 砷
arsenopyrite 砷黄铁矿
aryl 芳烃
ascend 上升
attainment 达到，成就
attenuation 弱小，薄弱，衰减，减弱，降低
attrition 摩擦，碎，研磨
augite 辉石
autogenous 不用焊料焊接的，自然的，自生的，自动的
autogenously 自动地

auxiliary 辅助的，帮助的，辅助部分

B

bacteria 细菌
bacterial 细菌的
baffle 障板，阻板，遮流板，反射板
ball mill 球磨机
ballast 碎石，平衡器，整流器
bands 带
barite 重晶石
barium 钡
barytite 重晶石
basalt 玄武岩，玄武岩制品
base metal 碱金属
basic rock 基性岩
basin 盆地，洼地
basket 篮，筐，岩芯管
batch 分批的，间歇式的，单元
battery 电池，电瓶
baumhauerite 褐硫砷铅矿
bauxite 铝土，矾土
be in question 成为问题，正在讨论
beach 海滩
beam 横梁
bearing 方向，方位
beater 打击者，锤子
bed 矿床
beeswax 蜂蜡
belt 皮带
bench-scale 单元试验
benjaminite 铜银铅铋矿
berndtite 辉铁锑矿
berryite 硒铜矿
berthierite 铌钛铀矿
berzelianite 硒铜矿
betafite 硫砷银矿
betekhtinite 自然铋
benzoic acid 苯酸，安息香酸
benzyl amine 苄胺
bias 偏差，位移，斜的，使偏
bicarbonate 碳酸氢盐
billingsleeyite 辉铋矿
binding 黏合的，黏合剂

biosphere 生物界圈，生物可生存的大气层，生命层
biotite 黑云母
bischofite 水氯镁石
bisulfate 硫酸氢盐，酸式
bite 咬住，夹住
bitumen 沥青
bituminous 沥青质的，烟煤
blade 刀刃，螺旋桨叶，刀片
blast 爆破，爆炸，鼓风
blinding 堵孔
block 堵塞
blow 打击，冲击，吹
boiler tube 蒸气管道、锅炉曾
bolt 用螺栓固定，螺旋
bomb 氧弹，弹
bomb calorimeter 弹式量热器
bond 结合，黏结，焊接
boot 罩，套管，料仓
boron 硼
boron hydride 氢化硼
boundary 边界，范围
bowl 罐，碗，斗，槽，转
brackish 有盐味的，稍咸
brass 黄铜制
breakdown 坏掉，细分，
brick-making 制砖
bridge 桥，架桥
brine 卤水，盐水
brine deposit 卤水矿
brittle 脆性
broken coal 破碎煤，碎块煤
bubble 气泡，泡
buffer 缓冲器，减震器
burden 负荷，负载
butyl 丁基
by-product 副产品

C

cabinet 小室，小间，盒，壳
cadmium 镉
cage 升降机，笼子
caisson 潜水箱，沉箱

calcareous 石灰质的，石灰的，含钙的
calibration 标准刻度
californium 锎，刻度的
calorific 生热的，发热的
cannel 烛煤
capacitance 电容量，机械产量
capacity 处理量，处理能力
capital 投资
capture 捕捉，捕获
carbinol 甲醇
carbonenrichcd 富含碳
carbonaceous 碳的，碳质的
carbonate precipimte 碳酸盐沉淀物
carbonization 碳化，焦化，干馏
carboxylate 使羧化，羧化物，羧酸盐
carrier 小车，矿车
cartridge 支架，座，滤筒，过滤器
cascade 串联的，格状物，阶流式，水柱，跌差
casting 铸件
catalyst 催化的
catalytically 催化地
cataract 水力制动机，缓冲器
categorize 分类，把……归类，区别
cationic 阳离子的
cell 浮选槽，格，筛网
cellulose 纤维素，细胞膜质，细胞的
centimeter 厘米
centrifugal 离心的
centrifuge 离心，离心机
centripetal 向心的，应用向心力的
charcoal 炭，木石炭
chain conyeyor 链条输送机
chalcocite 辉铜矿
chalcopyrite 黄铜矿
chalk 白垩，粉笔
choked crushing 阻塞破碎，拥挤破碎
chromate 铬酸盐，铬酸盐类
churn 搅拌器，搅拌
circuit 流程，工艺
circumfcrence 圆周，四周
clamp 使固定，夹子，压紧，支架，卡紧，定位
clarity 透明度，清澈度
classical 传统的，古典的，经典的

classification 分级
classifier 分级机
clasti 碎屑状，碎片
clay 黏土，泥土
clayey 含黏土的，黏土质的，泥质量的
clean coal 精煤
cleavage 劈裂（理），解理
cling 黏着，依附
clinker 溶渣，熟料，烧结，烧成渣块
clog 障碍物，障碍，堵塞
close circuit 闭路流程
closely 严格地，严密地，精密地，仔细地，靠近地
coagulant 凝结剂
coagulation 凝结，凝固
coal preparation 洗煤，选煤
coal preparation plant 洗选煤厂
coherent 黏着的，凝聚性的
coke 焦炭，炼焦
collapse 倒塌，消失，毁坏
collector 搜集器，捕收剂
colorimetrie 比色分析的
channel 槽钢，通道，管路，风沟，气沟，方法，手段
charmel sampling 刻槽取样
char 炭，碳桩柱一端
chart 略图，有刻度的记录纸
chemisorb 化学吸附
chemisorption 化学吸附作用
chipping 碎片，切，削，凿，碎裂，劈碎
chlorite 绿泥石
choke 堵塞，阻塞，拥挤
combustible 可燃性的，燃料
comm crcial eparation 工业分选
comminution 粉碎，粉（破，磨）碎过程，渐减
compatible 相容的，可共存的，一致的，协调的
complex 复合的，多元的
compressor 压缩机，压缩物
compromise 妥协，折中，平衡
connate 包含，暗示，指点
conyentional 传统的，常规的
concave 凹面的，中凹的
coneentrate 选矿，富集，浓缩物，精矿
coneentration 浓缩，富集

condense 冷凝，凝结，浓缩
conductance 传导性，导率，电导性
cone crusher 圆锥破碎机
conflicting 冲突的，矛盾的
conical 锥形的
consecutive 顺序的，相邻的，连续的
constraint 强迫，强制，约束
contradict 反对，驳斥，同……相矛盾
contrast 对比（度），对照，反差
conyex 凸面的，中凸的
conyeyor 输送机
conyeyor belts 皮带输送机
convincing 有说服力的，令人信服的
core 岩芯，晶内偏析，钻取土样
corrugated 波纹的，波形的，有槽的
countercurrent 逆流，逆电流的
crack 破裂，敲碎，裂开，裂缝
cresylic 甲酚，杂酚油的
crimp 皱缩，打褶，脆的，妨碍，障碍
criterion 准则，标准，判据
cross-sectional area 横断面积
crowd 许多，大量，群聚，积聚，拥挤
crucible 坩埚，炉缸
crucible swelling number 坩埚膨胀指数
crush 破碎
crasher 破碎机
crushing 破碎
crushing chamber 破碎腔，破碎室
crust 壳，表层，硬结，用外皮覆盖
crystal 结晶
crystalline 结晶的，晶状的，结晶体，水晶体
crystallinjty 结晶度，结晶性
cursory 草率的，仓促的，疏忽的
curvature 弯曲，屈曲，曲率，弧度
cushion 垫子，缓和……的冲击
cut-off 断开，关闭
cyanide 氰化物，用氰化物处理
cyclone 旋流器
cylindrical 圆柱的，筒形的

D

deactivate 使不活动，使无效
deactivation 减活化作用，惰性的

deek 筛面，筛网
decompose 分解，还原，风化
defloeculate 反絮凝剂，悬浮剂，解絮
deform 使变形
deformation 变形，形（应）变
degradation 降低，下降，退化
degrade 降低，退化
delay 退迟，延迟，抑制
deleterious 有害的，有害杂质的
deliberately 谨慎的，慎重的
delivery area 排料面积
demagnetizing coil 退磁线圈
demarcation 分界，划界，定界，区分
denominator 共同特征，命名者，分母
dellse 密的，密集的，稠的
dense liquid 黏稠矿浆
densifier 增浓剂，浓缩机
densimetric 密度的
density 密度
dcnsity composition 密度组成
depressz 压下，降低，减少，抑翩
depressant 抑制荆，抑浮剂
depression 减少，降低，沉降，抽空，排气，抑制
derivative thermog tavimetri 微商热重（DTG）
descend 下降，落下
designation 标示，指明，规定，选定
destructive 破坏的，有害的
detention 阻止，停滞，滞留
deteriorate 变坏，降低，恶化，损坏
eterioration 磨损
deviation 偏差，脱离，误差
dewater 脱水
dextrin 糊精
diagrammatically 用图解法，利用图表
dialkyl thiono carbamate 二烃基硫逐氨基甲酸盐
diaphragm 膜，隔膜，振动膜，隔膜的，阻隔
diaspore 硬水铝石
dichromate 重铬酸盐
dictate 命令，分配
differential thermal analysis 差热分析
differentiat 微分，求导数
diffusion 扩散，扩压，渗出，散布，传播
digestion 消化，吸收，溶解，加热浸提

dilatation 膨胀系数，扩张，传播
dilatometer 膨胀剂
dilute media 稀介质
dilution 稀释，冲淡，稀释之物
diminisht 减少，缩短，递减，缩小
disc 研磨盘
discard 放弃，抛弃，排出，除去
discernible 可识别的，辨别的
discharge 排料口
discharge ability 排料量
discrete 不连续的，分离的，分离合元件
disintegrate 使分离，剥离，碎裂
dispellse 节省去
disp ersant 分散剂
disperset 分配，散开，分散，粉碎，分散的
dispersing 分散的
disposal 处理，收拾
dissehainate 散布，浸染
dissipate 消散，散逸
distort 扭曲，使变形
distortion 变形，扭曲，失真，挠曲
distribution 分布（率）
dithiocarbamates 二硫代氨基甲酸
dithionate 连二硫酸盐
dithiophosphates 二硫代磷酸盐，黑药
diverr 使转向
dixanthogen 双黄原
dodeeylamine 十二烷胺
dolomite 白云石
downcurrent 下向流
drain 水管，下水道，排出，流出
drainaget 排水，疏水
draining screen 脱水筛
dressability 可选性
dual-frequency 双频
dump 翻车机，倾倒卸
duplicate 完全相同的，复写，复制，加倍
durability 耐久的东西，耐久性
dwell 留居，停留时间，详细研究，评论
dwell time 停留时间，停留阶段
dynamic equilibrium 动态平衡

E

earthrs sialic layer 地球的硅铝质层
eccentric 偏心轮，偏心的
eddy 漩流，涡流，回旋，旋转
effectiveness 效率
efficiency 效率，功效，性能
effluent 支流，废水，污水，排出物
ejector 竹，喷出物，射出物
elastic aterial 弹性物料
elasticity 弹性，伸缩性
electroform 电铸
electrode 电极，电极棒
electron probe microscope 电子探针显微镜
elevator 升降机，提升机
elimination 消除，消去，淘汰
elliptical 椭圆的，椭圆形
embarrassment 困窘，阻碍
emission 发出，散发，散发之物
emission spectrcscopy 发射光谱学
encapsulate 密封，封闭，压缩
encircle 环绕，包围
encouraging 鼓励，激励，促进
cndless 无穷尽的，永不停
cndothermic 吸热的，吸热反应，内热的
cndwise 直立，倒头，首尾相接，向着两端，在末端
energy dispersive analysis 能谱分析
enlarge 扩大，增大
entrain 携带，夹带，输送，产生
entrainment 夹杂
entrapment 俘获，夹住，截流，捕集
envisionz 想象，展望
epidote 绿帘石
episode （一系列事件中）一个事件，一段情节，插曲
equivalent 相等的
error area 误差面积
ether 醚，介质
evaluate 判断，评价，计算
evolution 开展，发育，进化论，开展的，发育的，进化的
excavate 挖掘，开挖，开采
exclusion 拒绝，排斥，排除在外的因素
exhaust 排气口，排出之废气，用尽，消耗
exit 出口

expand 使大，扩大，变大，膨胀
expandible 可延伸的，可膨胀的
exploration 勘探，探索，研究
exploratory 探索的，发掘的，调查的，研究的
exploratory drilling 勘探钻井
explosive 炸药，爆炸性的，易爆炸的，猛烈的
extra-hard 过硬的
extraction 抽取，萃取，提取
extraneous water 额外的，附加的，无关的

F

facet 面
facilitate 使便利（容易），便利，简化
false 错误的，不对的，不真实的
feature 外貌，特征（点）
feed chute 给料溜槽
feed table 给料盘（板）
feeder 给料机，给煤机
feedstock 原料
feed well 给料筒，给水井
feldspar 长石
ferrcsilicate 铁硅酸盐
fertilezer 化学肥料，人造肥料，肥料
fibre 纤维，构造，结构，性格
filler 填充物
filter 滤波器，口，透过，渗出
filtrate 过滤，滤液
filtration 过滤
finder 取景器，指导镜，定向器
fineness 细度
fixation 固定，决定
flaree photometric 火焰光度
flaw 破（断）裂，裂缝（隙），横断面
flexibility 灵活性
flexible 易弯曲的，柔韧的
flight 飞行，飞翔，逃亡，刮板
float 浮标，浮子，浮筒，使浮动
float-sink test 浮沉实验
floatable 可浮性
floc 絮凝物，絮片
flocculant 絮凝剂
flocculation 絮凝作用，絮状沉淀法
floor 铺，最低额，场地，打倒，完成

flotation 浮选，漂浮（性），浮悬浮
flowsheet 流程
fluctuation 变动，波动
flue 烟道，烟管，锅炉通气管
fluorescence 荧光性，有荧光性
fluxing agent 熔剂，焊药
flywheel 飞轮
forcibly 强行的，用力的
foremost 最初的，第一流的，主要的
format 版式，程序，格式，规格，形式
fouling 难闻的，肮脏的，堵塞的，污物，阻碍
foundation 基础，地基
fragile 易碎的，脆的，脆弱的
fragment 碎片（块），生成物，断片
frarhe 骨架，框架
framework 骨架，框架，梗概
free crushing 自由破碎
freshwater 淡水的，清水
friability 易碎的
friable 脆（性）的，易碎性，易剥落性
froth 泡沫，罐，起泡的
frother 起泡剂
function 指数，系数，幂，函数
fusion 熔化，碎，合成，汇合，聚变

G

gamma—ray emission Y射线发射（辐射）光谱
gangue 矿脉中的夹杂物，矸，脉石，尾矿，矿渣
gape 开口，张开（的阔度）
gasification 气化
gate 闸门
gauge 标准尺寸，量具（规，表），测量
gauze 金属，塑料网纱，铁纱，金属丝网
gear 齿轮
gelatine 明胶，凝胶
geological 地质学的
glue 骨胶
glycol 甘油
gono-go gauge 过端-不过端量规
goyazite 磷锶铝石
grade 品位
gradient 斜率，梯度变化曲线
granite 花岗岩

granular 小粒的，粒状的
graphical 图示的，图解的（=graphic）
graphite 石墨
graphitization 石墨化作用，处理，涂石墨
gravity concentration 重力分选
gray-king carbonization assay 葛金干馏试验
Greece 希腊
grindability 可磨性
grindability index 可磨性指数
grinding 磨碎，磨矿
grip 控制，支配，掌握
grizzly 格条筛
groove 凹槽
gross value 总值
guar 保证
gum 树脂，树胶，胶质，胶
gypsum 石膏
gyratory crusher 旋回式破碎机

H

half-cycle 半周期
halite 石盐
handle 搬运，处理
hazy 有薄雾的
heavy-duty 重型机械
hematite 赤铁矿
hemi-micelle 半胶束
hemisphere 半球活动的范围，领域，半球
heterogeneous 不均匀的，不同的，非均质的
heteropolar 具多极的，杂极性的
hindrance 阻碍，妨碍，干扰，延迟
histogram 直方图
hoist 提升
homogeneity 同类，同质
homogeneous 同类的，同性质的
homopolymer 均（聚合）物，同聚物
horsepower 马力
hose 软管，用软管输水洗
hoznbleude 角闪石
hub 中心，轮毂
humie 腐殖的
hutch 笼子，罐笼，跳汰机筛下室，通过跳汰机筛板的细料

hydrochloric 氢氯酸的，氯化氢的
hydrochloride 盐酸盐，盐酸化（合物），氢氧化（合物）
hydrocyclone 水力旋流器
hydrometallurgical 水冶的
hydtometer （液体）比重计，浮计，流速计
hydrophilic/hydrophobic 吸水性的，保持湿气的，吸湿的，亲水的
hydrous 含（结晶）水的，水合（化，状）的，含氢的
hydroxamate 羟肟酸盐
hydroxyl 氢氧化物
hygroscopic 吸水的，吸湿的，湿度计的
hygroscopic moisture 湿存水
hystcresis 磁滞（现象），滞后现象

I

idiocy 空转的，闲暇的
idling 空载，空转，无效功
ignite 点燃，发火
ignition 点燃，着火
illite 伊利石
illustrate 举例
immorse 浸入

L

liebigito 一铀钙石
lifter 提升装置
lignite 褐煤
lignitic 褐煤的
limb 肢，枝，翼，边，部分
lime 石灰
limestone 石灰石
limitation 限制，受限制，限制条件
limonite 褐铁矿
linear 长度的，线的，直线的
linen 亚麻，滤布，灰白色的
liner 衬（垫，里，板）
lining 衬
linkage 连结，联系，耦合，交联
linoleic 亚油酸
linolenic 亚麻的
lip 唇，堰

liquor 酒，酒类，液体，液
lixivant 浸出词（溶液），溶滤
lock-hopper 闭锁漏斗
log-log graph paper 双对数坐标纸
logarithmic 对数的
logging 电测，测井，记录
lorry 矿车，卡车，运料车
low-rank coal 低价煤
lump 使结块，成团，矿块

M

machine 机制，用机床加工
magnet 磁铁
magnetite 磁铁矿，氧化铁锈石
magn etizing block 磁极
maintenanc ecost 维修费
make up medium 补充介质
malfunction 发生故障
manganese 锰
manipulate 操作，处理
manometric 测压的，压力的，压差的，流体压力计
marble 大理石，矸石
marcasite 白铁矿
margin 边缘，余地
marine 海的，海生的
master 控制，征服
measure 复数，系，组
mechanical property 机械特性
medium (pt. media) 介质
medium conditioning and recovcry system 介质分配与回收系统
medium solids 加重质
megajoules 兆焦耳
melartterite 水绿矾
membranes 薄膜，膜片，振动皮光圈表层
mercaptobenzothiozote 巯基苯骈噻唑
mesh 网目，筛眼
metallo-organic （常用 metallorganic）有机金属的，金属有机物
methoxide 甲氧基
melhyl 甲烷基
mica 云母
micelle 胶束囊，胶质粒子

microscope 显微镜
microstructural 显微结构
mierowave attenuation 微波衰跌
middling 中等的，普通的
middlings 二等货，次等货
midway 中途的
migrate 迁移，移往
mill 选矿厂，工厂，研磨机
mill scale 轧钢铁屑
millis econds 毫秒
mineface 采煤工作面
mineral 矿物
mineral processing department 选矿部门
mineral processing plant 选矿厂
minefalogical 矿物学的
mirabilite 芒硝
mirror-image 镜像
miscellaneous 有不同性质的，各式各样的，多方面的，混杂的
miscible 可混的，互溶的
misleading 错配物，错误地引入歧途
mispickel 毒砂，砷黄铁矿
misplaced material 错配物
mixing 混合
mobility 松散（度），可（活，流）动性，迁移率，淌度
mode 方式，模式
model 仿照（造），模型
moderate 有节制的，有合理限度的，节制，缓和，减轻
modifier 调节剂，调节器
mogensen sizer 摩根生筛，概率筛
molybdate 钼酸盐
monomer 单（分子物）体，单基聚物，单元结构
monoxide 一氧化物
montmorillonite 蒙脱石
motivate 促使，为……的动机
moulded specimen 样品
moving sereen 动筛
muffle furnace 马弗炉
mullite 富铝红柱石，莫来石
multideck screen 多层筛
multiplicity 多，大量

muscovite 白云母

N

National Coal Board 国家煤炭管理局
nearmes particl 难筛粒
near-gravity material 难选矿物，密度物
near-size particle 难筛粒
negligible 可忽略的，不计的，微不足道的
nest 组，套，窝
net 纯净的，最后的
net return 纯利润
neutralization 中性，不动
neutron activation 中子激活
neutron capture 中子俘获
nil 零
nip 挤，挟，咬
nodule 结核
nominal 标称的，名义上的
non-banded 不带条纹的，不连接的
non-coredhole 无核孔洞
noncorrosive 无（非）腐蚀性的，不锈的，抗腐蚀的
noncombustible 不可燃的
nonnoatability 不可浮性
nonmagnetic 非磁性的
nonmicrcscopic 非显微的
nontoxic 无毒的
nonvolatile 永久的，长存的，不挥发的，非挥发性的
normal to 垂直于……
normative （定）标准的，规范的，正常的
nuclear magnetic resonance 核磁共振（常用 nuclear magnetism resonahce）
nucleonic 核（电）子的，核物理的

O

obstruction 阻塞，妨碍
obviate 消除，排除，预防
obviated 有预防措施的
octadecylamine 十八烷胺
offset 抵消，弥补
open area 开孔面积
open circuit 开路流程

optical adsorption spectrophotomte 谱分析仪
optical emission 光学发射
optical microscope 光学显微镜
ordinate 纵坐标
ore 矿石
ore-processing department 选矿部门
orientation 取向，方向，定向
oriffice 小孔，小洞，结流孔，阻隔，调整光圈
origin 起源，由来，起点，原因，来历
orthoclase 正长石
osmotic 渗透压，压力，压强
outcrop 露头
outgrowth 副产品生长物结果
output 生产能力，效率，流量
over 盖上，放在…上面
overburd enstrata 过负荷
overflow 溢出，溢流，溢流管
overgrindi 过粉碎
overhead 高高的
overlie 放在……上面
overlying 过分地，过度地，太，上覆的，叠加的
overlying deposit 表沉积层
overlying strata 上覆地层
oxyhydryl 羟基

P

pack 充填，压紧
packing 充填物，填料
pad 垫片，填料
paddle 浆状物，闸门，开关
paddle whcel 浆拌叶轮
pan 盘
parabolic 抛物线（性）的
paraffin 石蜡，链烷烃，石蜡油，煤油
parallelepiped 平行六面体的
parameter 参数，系统，特性
parformance 技术（操作）性能
particle size 粒度
partings 分离的，夹层的
partition coefficicnt 分配率
partition curve 分配曲线
partition error 分配误差
pawl 棘爪（轮），卡子，凸瓣，倒齿

peat 泥煤，泥炭
peel off 剥落
penctrate 刺穿，透入，凿穿
percolated 秒，渗透
percolated through 穿过
perforation 穿孔，打眼，多孔冲裁
performance 钟，性能，特性，效能，效率，生产力
peridotite 部，橄榄岩
periodic 周期的，定期的，间歇的
peripheral 圆周的，周边的
permanganate 高锰酸盐
permeability 渗透（性，度），穿透性
permo-triassic 二叠纪-三叠纪
perpendiculzr 垂直的，正交的
petrogaphie 岩石的，岩相的
Petrographic(al) 岩石（学）的，岩相（学）的
Petrographic(al) thin-section 薄片
Perchloroethylene 全氯乙烯
phqsphate 磷酸盐
pick up 拾（起，取），挑选，抽出，捡拾
pillar （支，煤，矿）柱，柱状物
pillar samptig 煤柱取样
pillow 轴衬
pilot plant 中试厂，试验厂
pine 松木
pioneef 倡导者，发明，创造
pitch 间距，节距，齿节
pitman 连杆
pivoted 装在枢轴上的
pivot 枢轴，支点
plagioclase 斜长石
plant （整套）设备，厂，车间，厂矿
plastometer 塑性（度）计
pliers 钳子，夹钳，扁嘴钳
plot against 画出……对……的曲线
plot 绘图，画曲线，作图表示
plummer block 止推轴承
plunger 活塞，滑阀
ply 层，胶，厚度，叠加，倾向
pneumatic 气功的，气力的，气胎
polished blocks 抛光片
pollutant 污染的，污染物

pollutants 工业污染物质
polyaerylamide 聚丙烯酰胺
polyelectrolyle 聚合物电解质
polyethylene 聚乙烯
polymer 聚合物，多聚物
polymeric 聚合的
polypropytcnc 聚丙烯
polyvalent 多价
pond 池，塘，坑，堵水或池
porosity 多孔性，气孔率，松散度
porous 多孔的，疏松的
portray 描绘，描述
positive 强制的，正的，肯定的
posslbllltv 可能性，希望，可能发生的事
postulaced 假定，主张，要求，以…为前提
powder 粉末，矿粉
preeaution 预防，保护
preeede 位于……之前
preliminary 预备的，初步的
presized 预先分级的
presizing 预先筛分
pressure-resistant metal 耐压金属
pressurize 增压，产生压力，密封
pretreatment 粗加工，预先处理
prewet 预先湿润，预湿
prewetted 预先润滑的
primary acparation 粗选
primary crushing 初碎，粗碎
probability 可能性，概率
probabilistlc 随机的
probability 概率，几率
probable 可能的
probable error 可能偏差
process 工艺（过程，方法）
profile 轮廓，外形，剖面
pronounced 明显的，显著的
proponent 支持者，建议的，支持的
protrude （使）伸（突）出
protruding 伸出，推出，突出
provision 准备，（预防）措施，构造
pulley 皮带轮
pulsation 脉动
pulveizc 磨碎，研磨

pump 水泵,(用泵)抽
punch 冲孔,穿孔
pycnometer 比重瓶(管,计),比色计
pyramid 金字塔,矩形,聚成一堆
pyrite 黄铁矿
pyritic 黄铁矿的
pytrhotite 磁黄铁矿

Q

quadruple 四倍地
quantification 容量,量化,以数量表示,定量
quarfy 采石场,采石
quebracho 树木,百雀树

R

rabble 搅拌棍,搅拌
radiate 辐射
radloisotope 放射性同位素
ragging 重粒料铺层,跳汰机人工床层
rake 倾伏,斜脉,耙
rank 等级,级别
ratchet 棘轮,齿杆
raw coal 原煤
rayon 人造纤维
reagent 药剂,试剂,反应物
reasonable 适当的,比较好的,合理的
recalking 重凿缝
recede 退回,后退
receding 后退的
receiving area 受斜面积
recess 切口,凹口,凹进
recirculation 循环
recirculate 再循环,回流
reclaim 回收,再生,再处理
reclean 再洗
recover 恢复,回收,再生
recoverable 可回收的,可找回的
rectangular bock 矩形块
recycle 再重复,反复循环,回收,重复利用
reduce 减少,压缩,还原
reduction 减少,降低,还原作用
refer 把…归类,起源于,提及,委托,参考,引证

refinement 精加工
reflux 侧流,不淌,加热
refractory 耐溶的,高熔点的,难辨的,不易处理的,耐火材料
refuse 熔化,废物,渣溃,无用的,矸石
regime 状况,状态,方式,方法,领域
reject 拒绝,排斥,报废,抑制,干扰,矸石
reliable 可靠的,安全的,确实的
reliance 信赖,依靠,信心
remedy 补救,校正
render 重发,再生,提出,执行,给予,使……变得,缴纳
replicate 复制品
report(to) 进入,报告
resembling 相似的,类似的
residue 剩余物,残余物,残渣,沉积物
resistance 抑抗,阻力
resonance 共振
resonance screen 共振筛
restrict 限制,约束
retard 延迟,阻止
retention 停留
retort 容器,蒸馏,提纯
retort tube 蒸馏试管
reverse 颠倒(的),相反(的),倒退(的)
rewashing 再洗
reweigh 再称重
ribbon 带子,带状物
riffle 格条,格槽缩样器
rigorously 严格地,精确地
rinsing screen 脱介筛
rip 撕(开,裂),劈(凿)开,裂开
riser 提升器,上升装置,立管,整管,溢流口
road-making 筑路
rod mill 棒磨机
roof 顶部,绝对上升限度,覆盖
rosin 树脂,用树脂擦
rotary breaker 旋转破碎机,选择性破碎机
rotating probability screen 旋转概率筛
rotor 旋转器,滚筒,转子
rough 粗的,不平的,粗制
route to 按规定路线送
routine 常规的

routine analysis　常规分析
run-of-mine　原矿
run-of-mine ore　毛矿
rutile　金红石

S

saline　含盐的，咸的
salinity　盐（浓度）度，咸度，含盐量
sample　样品，试样
sampling　抽样的，抽样检查
sanidine　透长石
sapropelic coal　腐泥煤
saran　砂纶，萨冉树脂
saurolite　十字石
scalar　纯量的，数量的，标量的
scalp　筛，脱泥，选前筛出大块
scalping screen　脱泥筛，原煤分级筛，筛除大块的筛子
scanning election microscope　扫描电子显微镜
scheme　流程图
scheelite　白钨矿，重石
scraper　铲运机
screen　筛子，筛分
scrubber　洗净机，煤气，洗涤器
seam　缝，煤层，矿层
secondary crushing　二次操作
secondary separation　二段分离再选
sectionalize　分段（节，组）
sediment　沉积物，沉淀物，残渣
sedimentary　沉积岩，水成岩
seek　找，探，寻
seepage　渗漏，渗出
segment　部分，段，节，分裂，切断，扇形体
segregate　分离，隔离，离析
segregation　分离，离析
self-feeding　自入料，自动给料
semi-logarlthmic　半对数的
semistable　平稳的
separation　分离，分选
separator　分选机
settling　沉降
settling rate　沉降速度
shaft　轴，（竖，通风）井

shale　油页岩
shalter　打（破，击）碎，破（碎，裂）片
shear　剪切，切，剪（力，切），剪应变
shell　外壳
shift　（调）班，轮班（制），移动
shrink　收缩，缩小
shut　关闭
shutdown　关闭，断路，非工作周期
sialic　硅铝质
siderite　菱铁矿，陨铁，蓝石英
sieve　（细）筛，滤网，筛分，滤
seve analysis　小筛分分析
sieve bend　弧形筛
silk　丝织物，丝状的，丝制的
simplification　简化
siphon　虹吸
size analysis　筛分分析
size consist　粒度组成
size fraction　粒级，粒度级别，筛序
size reduction　（使）破裂（碎），（使）断裂
skew　斜的，歪的，非对称的，误用的
skim　垫片，去沫，脱脂
skinuner　（泡沫）分离器
skin　表面，壳，外板
skip　箕斗
siabby　板（片，层）状的
slag　渣，火山灰岩，结液，排法
sleeve　套（筒，管，垫）
slim　细的，微弱的
slime　（黏，煤，矿）泥，微粒
slip　滑动，滑行，滑掉
slippery　润滑的，光滑的
slope　斜度（米），坡度
slowed down to　把速度降低到
sludge　煤泥，泥泞，污水，沉积物
slug　猛击，嵌条，床条
sluice　水闸门，溢水道，溜槽
slump　坍塌，滑动
slurry　矿浆，煤浆，泥浆
smelter　熔炉，冶金厂
soda　苏打，氢氧化钠
sol　溶液，溶胶（soltion）
solenoid　螺线管，线圈

sophisticated 使复杂化
span 间距，跨过，跨越
sparking 火花，闪光，激发，鼓舞
species 形式，外形，物质，种类
specific gravity 密度
specification 技术规格，（规格）说明书，详细说明，技术规格
spectroch emical 光谱化学的
spectrometric 摄谱仪的
spectrometry 光谱测定法
spectrophotometric 分光度（计）的
speeded up 加速，增速
spider 支座，臂架
spindle 铜杆，轴，心轴，柱，杆，星形，机（支）架
spiral 螺旋的，螺旋线的，螺旋分选机，螺旋，螺旋形
spiral off 螺旋刮板
split 裂，分，劈，裂开的，分裂的，剖分的，分裂，离开
split off 使分裂出来，使分离
spore 苞子，芽苞，胚种，种子
stable 稳定的，安定的，紧固的，不变的
stack 堆，组，套，堆积，套叠，堆满
stack gases 烟气
stack gases scrubber 洗涤器，除尘器，净气器
stagger 交错，错列，排列，交错的，错开的
stainless steel 不锈钢
standpipe 圆桶形水塔，立管
starch 淀粉，浆，给……上浆
stationary screen 固定筛
stator 定子，定片，导叶
steam raising 蒸汽蒸发，发气
steepness 陡度
stem 杆，棒，柄，把，发生于，起源于
stiffncss 刚性，韧性
stocspile 堆，存货，储存
storage 储存
strain 拉紧，紧张，变形，损坏，弯曲，过滤
strati-tcation 层，成层，分层取样
stress 应变，变形
stretch 伸，拉，弹性的，展开，扩大
strike 打击，撞，冲击

stringent 严格的，严厉的，迫切
strip 剥取，摘取，带条，窄条，剥，沿断层开采
stroke 打（冲）击，冲程
stub 树桩，短桩，剩余部分，节，抽头
sub-bituminous coal 次烟煤
sub-micron-size 超细粒
submerge 浸没，浸在水中，潜水
submergence 浸没，没入水中
subshell 亚层，子壳层
subtraction 减去（小，结）扣除
suction 吸，空吸，抽气
sulfidation 硫化作用
sulfide 硫化物，硫醚
sulfonate 硫酸化，硫化
sulphide 硫化物
sump 溶液槽，排水坑，水池，水仓
superscde 代（接）替，取（答）代，更换（迭），废除（弃）
superstructure 上层结构，上层建筑，超结构
surfactant 表面活性剂
surge bin 缓冲仓
suspersion 悬、吊、挂、悬浮物，悬胶液
sustain 支撑，继续证实，确认
sustenanee 粮食，维持，营养，支持
swamp 沼泽，湿地，煤层聚水洼，淹没，搪塞
swell 膨胀，胀大，时髦的
swelling number 膨胀系数
swift 快的，迅速的，急流，激流
swing 摆动
sylvite 钾石盐
symjnetrically 对称地，平衡的
syngenetic 同出的，共成的
sysiematie 系统的，体系上，有次序的，特直的，有计划的，整齐的

T

table 桌子，放在桌子上，拔床
tableing 嵌合，招床选矿，制表，造册
tabulate 把……列成表
taconite 铁磁石，铁英石
tailing 尾矿
tailored 简单明了的，简洁的，特别的
tank 容器，箱

tannin 鞣酸（类物），丹宁
taper 圆锥、尖锥，锥形的，弄尖，（使）逐渐变细
tar 焦油，柏油，涂柏油，（涂有）焦油的
target 目标，指标
teat 突出物，凸起，接头，轴颈
tedious 冗长的，乏味的，慢的
teeter 摇摆不定的
tenacity 坚韧，弹性，抗断强度，坚持
tensile 拉（张）力的
tension 张力，紧张，拉伸
ternary 三个的，三重的
tertiary 细碎，三次破碎
tetrabromoethane 四溴乙烷
thermal neutron 热中子，热能电子
thermistor 热敏电阻，热控管，半导体
thermochemical 热化学的
thermogravimetric analysis 热重分析
thickener 浓缩器，浓缩机，沉降槽
thin 使变薄
thiocarbanate 硫代碳酸盐
thiocarbsnilide 硫黄
thiol 硫醇类
thorough 彻底的，详尽的，根本的
thronle 节流阀，截流阀门，结流，扼流
throughput 生产量，流量、产量，处理量
throw 扔，投，冲程，摆幅
tile 瓦，砖
titration 滴定法
titrimetrieally 滴定的
titrmetric 滴定的
toggle 肘（板）
tolerance 公差要求
toluene 甲苯
top size 粒度上限
torque 转矩，扭矩，扭转
tough 坚韧的，不易磨损的
tourmaline 电气石
trajectory 弹道，轨道
tramp 错配物
transition 转变，转化
trap 陷阱，圈套，收集器，随身携带的，截聚
treatise （专题）论文
trench 沟，壕，挖沟
trickly 一滴一滴流的
trouble-free 无故障的，安全的，可掌的
trough 槽盆，槽形的，开槽（沟）
tumbler 大玻璃杯齿轮，开关，滚筒
tumbling 转鼓，限转
tyre 轮箍

U

ultimate 最后的，结局的，极限，终端
ultlmate analysis 元素分析
ultrafine 特细的
ultrafine 超声（波）的，超声速的，超声波
ulbite 方钛铁矿
ullmannte 锑硫镍矿
umangite 红硒铜矿
unavoidable 不可避免的，不得已的，不能废除的
unblind 解放，解开，未被堵塞的
undertake 着手作，进行
undergo 经历，经受
undergone 经历，经受，受到
undergrinding 磨的不够
undersize 尺寸过小，筛下斜，筛下，尺寸不足的，小型的
unduly 过分的，不相称的
unequivocally 不含糊地，明确地
unhindered 无阻碍的，不受限制
unified basis 统一的基础
unit 设备
unit operation 单元操作
universal crusher 通用破碎机
unlocked 分开，释放，解离
unobstruet 无阻碍的，没有阻挡的，自由的
unrestricte 无约束的
uphill 倾斜的
upward current 上升流
uraninite 晶质铀矿
uramothorite 铀钍矿
ustarasite 柱硫铋铅矿

V

vacuum 真空
vaesite 方硫镍矿

valleriite 墨铜矿
vanner 陶矿机，陶选带
variability 变化性，变率
veetorially 矢量地，向量地
verbal 口头的，非书面的
versatile 通用的，万能的，多用途的，可变的
versatility 多功能的，变化性
vesicle 泡，气孔，小穴
vesicular 多孔（状）的，小泡的，蜂窝状的
vessel 容器
ventilate 通风，排风，流滩，开气孔
veenite 维硫锑铅矿
via 经过，通过，借助于，道路
vibrating screen 振动筛
villamaninite 维拉曼矿
violarite 紫硫镍矿
violent 强烈的，极端的
viscometer 黏度计
viscosity 黏性，黏度
visualize 目测，观察，具体化
vital 必须的，主要的，极度的，要害的，有生命力的，有生机的
volatile 挥发的，易挥发的，挥发性物质
voltzite 肝锌矿
volynynskite 沃仑斯基矿
vonsenite 硼铁矿
vortex 涡流，蜗旋，旋转
vortex finder 溢流管
vrbaite 硫砷锑铊矿
vredenburgite 磁锰铁矿
vulcanite 软碲铜矿

vysotskite 硫镍钯矿

W

wadeite 钾钙板锆石
wardite 水磷铝钠石
water opal 玉滴石
wikeite 硅硫磷辉石
wischnewite 硫酸钙霞石
wodginite 锡锰
wollastonite 硅灰石

X

xanthophyllite 绿脆云母
xenotime 磷钇石
xomotlite 硬硅钙石

Y

yttrocerite 铈钇矿
yttroparisite 钇氟菱钙铈矿
yugawaralite 汤河原石

Z

zasper 碧玉
zeunerite 翠砷铜铀矿
zinnwaldite 铁锂云母
zircon 锆英石
zircophyllite 钙钛锆石
zircophllite 锆叶石
zirkelite 钙钛锆石
zoisite 黝帘石